“上海暑期学校——中华餐饮文化”项目指定教材

双语版

Basic Course of
Chinese Culinary Culture

张田田◎主　编
马思鸣　孙晨阳◎副主编

中华餐饮文化教程

基础篇

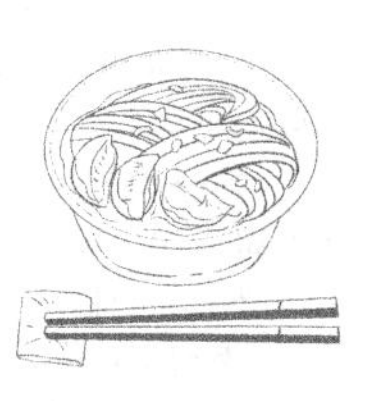

上海财经大学出版社

图书在版编目(CIP)数据

中华餐饮文化教程. 基础篇：汉、英／张田田主编.
—上海：上海财经大学出版社，2020.10
ISBN 978-7-5642-3532-1/F·3532

Ⅰ.①中… Ⅱ.①张… Ⅲ.①饮食—文化—中国—教材—汉、英 Ⅳ.①TS971.2

中国版本图书馆 CIP 数据核字(2020)第 077383 号

□ 责任编辑 肖 蕾
□ 书籍设计 张克瑶

中华餐饮文化教程：基础篇
（双语版）

张田田 主编
马思鸣 孙晨阳 副主编

上海财经大学出版社出版发行
（上海市中山北一路 369 号 邮编 200083）
网 址：http://www.sufep.com
电子邮箱：webmaster@sufep.com
全国新华书店经销
江苏凤凰数码印务有限公司印刷装订
2020 年 10 月第 1 版 2020 年 10 月第 1 次印刷

710mm×1000mm 1/16 20.25 印张 297 千字
定价：78.00 元

#【序】

《中华餐饮文化教程：基础篇》（双语版）终于面世了，尽管只是“基础篇”，但相信陆续会有系列教材及衍生作品与广大读者见面。该书是上海第二工业大学国际交流学院汉语国际教育团队的成果，而作为学院曲折成长的见证人和参与者有幸为之作序，这实在是一件快事。

《中华餐饮文化教程：基础篇》（双语版）得以出版，首先归功于上海市教育委员会打造的上海教育国际合作交流品牌“上海暑期学校”项目（Shanghai Summer School，简称“3S”项目）。2015 年，上海第二工业大学与中华职校联手运作韩国安山大学“中华餐饮文化”短期项目并获得成功，取得良好的社会效应，此项目也进入了市教委领导视野，机缘巧合地于次年被增补为“3S”项目。自 2016 年以来，“上海暑期学校——中华餐饮文化”项目连年举办，受到在校各国学员和社会各界好评，也直接催生了《中华餐饮文化教程：基础篇》（双语版）。在此，应该感谢上海市教委杨伟人先生、葛静怡女士的不吝指教，感谢姜丹女士的牵线搭桥，感谢中华职校黄玉瑾、薛计勇、林晨旭等领导的不懈支持，感谢上海市餐饮烹饪行业协会王慧敏会长、沈思明老会长的备至关怀，感谢静安洲际酒店以及静安英迪格酒店总经理嵇东明先生、马勒别墅酒店刘洪大师以及衡山集团、洲际集团旗下多家高星级酒店专业人士的不离不弃，感谢宋宝儒、俞涛、莫亮金、谢华清、徐余法等校领导自 2015 年以来对该项目的站场助威，感谢每一年“3S”项目的各位特邀师资与专家、项目班主任及中外学生志愿者的付出。有了你们，才有“上海暑期学校——中华餐饮文化”项目的延续和《中华餐饮文化教程：基础篇》（双语版）的问世！

从2015年项目实施之初，我们就意识到应该编写一本适用于该项目的教材，既可以规范项目内容、不断提升项目质量，也可以帮助学员及所有对中华餐饮文化怀有兴趣的中外人士更加全面地感知中华文化的博大精深，尤其是不受地域和时间限制随时研读中华餐饮文化读本，还可以凝结汉语国际教育团队成员的学术积累及教学心得。经过三年多的实践，该教材终于以汉英双语、图文并茂的形式与大家见面；而且在教材编写期间，由汉语国际教育团队历经四年打磨雕琢的"汉语国际教育"本科专业也在校领导关怀下成功获得教育部批准备案，已招收首届2019级中外新生，不仅使该教材有了专业级新读者，也即将拥有一批未来的导读者、诠释者。

祝愿《中华餐饮文化教程：基础篇》（双语版）和"上海暑期学校——中华餐饮文化"项目相得益彰，而且有后续佳作源源不断。

祝愿本教材编写团队和汉语国际教育专业未来有更好的发展。

期待中华餐饮文化不断发扬光大，成为一张最靓丽的中国名片，成为"一带一路"倡议不可或缺的百花园，而《中华餐饮文化教程：基础篇》（双语版）则是一张朴实无华的入场券。

如此，足矣！

孟昭上

【前言】

中华饮食文化博大精深、源远流长、内涵丰富，以其精湛的工艺、丰富的菜品、完整的流程、深厚的底蕴在世界上享有很高的声誉。关于中华饮食文化的专著和教材数量虽不少，但是针对外国留学生这一读者群体的却不多。编者通过上海市教委搭建的"上海暑期学校——中华餐饮文化"项目这一平台，结合多年来留学生教育及汉语教学实践，编写《中华餐饮文化教程：基础篇》（双语版），努力以通俗易懂的文字和丰富生动的图片来展示中华饮食文化的若干重要方面。

本教材的主要读者对象是热爱中华餐饮文化的外国留学生和其他中外友好人士，同时也适合中国学生和广大饮食文化爱好者。本教材在编纂原则上突出系统性、趣味性和通俗性。《中华餐饮文化教程：基础篇》（双语版）主要涵盖六个方面的内容：日常饮食、节庆饮食、地方菜系、特色小吃、茶酒文化和筵宴文化。"日常饮食"主要从饮食结构、烹饪方式、饮食惯制和饮食工具四个部分展开；"节庆饮食"则侧重于介绍中国传统节日和节日必备食物；"地方菜系"主要介绍中国现有的八大菜系及其代表菜；"特色小吃"则从小吃的角度让大家了解中华饮食文化的博大精深；"茶酒文化"和"筵宴文化"也是中华饮食文化的重要组成部分。

本教材第一章由孙晨阳编写、第二章由姚若冰编写、第三章由张田田编写、第四章由柳承吟编写、第五章由童欣婕编写、第六章由马思鸣编写，全书由张田田统稿。由于编者水平有限，本书疏漏之处，敬请读者批评指正！

编　者

2020 年 1 月 10 日

【总目录】

中华餐饮文化教程

基础篇

#【目录】

第一章　日常饮食

饮食是人们日常生活中最基本的、必不可少的需求。俗语说：“人是铁，饭是钢，一顿不吃，饿得慌。”从更深远的意义上说，饮食是人类存在与发展的基本前提，正如马克思所言：“人们为了能够‘创造历史’，必须能够生活。但是为了能够生活，首先就需要衣、食、住以及其他东西。因此第一个历史活动就是生产满足这些需要的资料，即生产物质生活本身。”① 自古以来，中国人对饮食的重要性都有明确的认识，孔子有言：“食、色，性也。”《礼记·礼运》有言：“饮食男女，人之大欲存焉。”吃东西是人类最基本的属性之一。随着历史的发展也形成了中华民族独特的饮食文化，孙中山曾经说：“我中国近代文明进化，事事皆落人之后，惟饮食一道之进步，至今为文明各国所不及。”②

一、饮食结构

对中华民族日常饮食结构的描述，可以从纵向、横向两个角度来展开。纵向角度是历时性角度，由于民族的迁徙、生产力发展、物种的传播，每个历史

---

① 《马克思恩格斯选集》第一卷［M］. 北京：人民出版社，1972：79.

② 孙中山著，牧之等选注. 建国方略［M］. 沈阳：辽宁人民出版社，1994：5.

时期中华民族的饮食结构都各有特点。横向角度是共时性角度，因为地理环境、气候条件、民族成分等的不同，所以在中国幅员辽阔的疆域里形成了不同的饮食文化区域。

（一） 饮食结构的历代演变

从历时性角度出发，有学者将中华饮食分为史前时期、先秦时期、秦汉时期、魏晋南北朝时期、隋唐五代时期、宋辽金元时期、明清时期和近现代时期，在各个时期因为生产力、农作物种类等不同，所以饮食结构也各不相同。

1. 史前时期

“茹毛饮血”是史前时期饮食的基本特征，是指原始人连毛带血地生吃禽兽。所谓“毛”，并不是现在意义上的毛皮之毛，而是指蔬菜类的植物。野生植物的采集和渔猎是史前时期食物来源的两种主要方式。

野生植物的采集是史前时期食物来源最重要的方式，因为渔猎在当时不能稳定地提供食物。在蛮荒之地中长期的生活实践也使人们可以辨别哪些植物可以食用，哪些是有毒的不可以吃，关于神农氏的古老传说就是对这一过程的反映。《淮南子·修务训》记载，神农氏“尝百草之滋味，水泉之甘苦，令民知所避就，当此之时，一日而遇七十毒”，展现了在史前时期，先民们为饮食所付出的牺牲和取得的成就。现代云南少数民族传统古歌中对采集野生植物作为食物的情形也有明确记述：“爬上高山，摘背野果回家来，剥剥过一天。进到深箐，找把野菜回家来，煮煮吃一顿。扛一捆栗柴回家来，烧一塘旺旺大火，双脚弯弯过一夜。”[①] 从目前考古发掘来看，有大量采集食物的遗存。如距今8 000年左右的湖南澧县八十垱遗址，发现了大量的菱角遗存。裴安平先生认为：“如将它们折合成食物量，当绝不亚于已发现的稻谷与稻米。”[②] 在中国史前遗址中发现的野生植物主要有：菱角、麻子、野生稻、槐树子、栗、梅、杏

---

① 刘怡、芮鸿．活在丛林山水间：云南民族采集渔猎［M］．昆明：云南教育出版社，2000：15.

② 裴安平．彭头山文化的稻作遗存与中国史前稻作农业再论［J］．农业考古，1998（1）：197.

梅、杏、李、樱桃、桃、柿、酸枣、榆钱、核桃、榛子、松子、梨、山楂、甜瓜、大豆、橄榄等。

尽管捕猎动物是比较困难的，但是渔猎也是史前时期重要的食物来源之一。随着生产力的提高，捕猎的收获也逐渐增多。渔猎起源很早，在陕西蓝田人遗址中，就已经发现有用于捕猎的石球，之后的陕西梁山、河南三门峡等遗址中也有大量发现。到了旧石器时代晚期人们又发明了弓箭，在距今30 000年的山西朔县遗址中发现石镞，说明当时人们的狩猎技术已经有了较大的发展。

从当前中国境内的考古发掘来看，渔猎的收获主要有以下几类：一是鹿类、野猪、野牛等动物。北京山顶洞人的遗址中有大量鹿类动物骨骼的遗存，与山顶洞人同期的安徽和县龙潭遗址中有大量斑鹿、牛、猪、虎、象等动物骨骼的遗存。浙江嘉兴马家浜文化遗址发现："在T1、T2的50平方米中，约有兽骨1 000千克，特别是在下层的底部，其厚约20~30厘米全为兽骨堆积。"① 二是鱼、蚌等水生动物，这些主要集中在水资源比较丰富的地区。如浙江河姆渡文化遗址中存在大量此类遗物："在遗址发掘中，鱼类、龟鳖类、蚌类等水生动物之多，不胜细数，个体更无法全部区分统计。我们清理了很少一部分龟的遗骸，明显地可分为龟类个体的就有两千多件。鳖类数量也相当可观。蚌类因受溶蚀，在现场只见一片晶亮的白色物体分布于地面上。还能看到许多被烧破了的陶釜中装有鱼类、龟鳖类、蚌类等水生动物遗骸。连狗粪中也可以清楚地看到鱼骨等碎片。"② 三是昆虫类，比如蚕蛹、蚂蚁、蚂蚱等，因为这些食物难以保存，所以目前考古发现比较少。但是从人类学的研究角度来看，这些富含蛋白质的食物应该也是史前时期人类食物中重要的组成部分。

随着植物培育和动物的驯化，农耕与畜牧逐渐代替了采集和渔猎，成为获取食物的重要来源。中国早期种植的农作物主要有：粟、黍、稻、麻、麦等。

粟、黍、麦是长江以北地区典型的农作物，其中粟、黍是起源于中国的农

---

① 浙江省文物管理委员会. 浙江嘉兴马家浜新石器时代遗址的发掘［J］. 考古，1961（7）：351.

② 浙江省文物考古研究所. 河姆渡：新石器时代遗址考古发掘报告［M］. 北京：文物出版社，2003：198.

作物，它们生长期较短，比较耐旱，适合黄河流域的气候，因此广泛种植于黄河流域，是当时人们的主要食物。麦是外来物种，大概是在公元前 5 000 年至公元前 3 000 年从西亚传入。但是由于早期食物加工的方式主要是煮，麦粒不经脱皮处理，难以煮熟，且食用后容易肚胀，因此在发明磨之前，麦的食用并不广泛。稻是起源于中国的农作物，目前早期出土稻的遗址集中在长江中下游流域，距今 6 000~7 000 年的河姆渡遗址出土的数量最多。麻在史前时期主要是纺织品的来源，而且麻子可以食用，河南和甘肃等地的考古遗址中也都发现有大麻子，说明当时麻子也是食物的一种。

猪、狗、羊、牛、鸡等动物是较早被成功驯化的，也成为中华饮食中重要的组成部分。猪是中国史前遗址中比较常见的家畜遗存，距今 1 万年左右的河北徐水南庄头遗址出土过猪骨，南庄头遗址的发掘报告认为所出土猪骨有可能是家猪。狗、羊、牛、鸡的遗存也曾出现，但是数量都不是很多。

2. 先秦时期

进入夏商时代，随着耕作工具、方式的发展，劳动力的增加，农业种植基本上代替了采集，成为食物生产的主要方式。畜牧业也有了较快的发展，圈养的方式已经普及，养殖种类和产量都有增加。

先秦时期主要的粮食作物有粟、黍、麦（小麦）、牟（大麦）、稻、菽（大豆）和麻等，基本种类与史前时期相似，但种植区域和产量却大为增加，这主要得益于金属生产工具的使用和普及，耦耕、精耕细作等方式的采用也增加了土地的利用率和农作物的产量。这一时期麦、菽的种植面积较以前有所扩大。从区域来看，北方地区主要以粟、黍、麦、菽为主，南方地区以稻米为主。麦的种植量的增长，主要得益于磨的发明与发展。

商朝畜牧业发达，有人认为商朝是游牧民族建立的，所以从商汤至盘庚时期国都经历过七次迁移。从目前甲骨文的记载来看，有刍、牧、牢等体现牲畜圈养的词汇，表明当时畜牧业已经发展到了一定的程度。《管子·轻重戊》写道：“殷人之王，立皂牢，服牛马，以为民利。”这说的就是商代畜牧业的发展状况。《管子·立政》又说：“六畜育于家，瓜瓠荤菜百果备具，国之富也。”

其中，六畜指的就是马、牛、羊、鸡、犬、猪。马、牛主要供劳动力使用，羊、鸡、犬、猪多作为食物。

在日常饮食中，士大夫阶层可能食肉类比较多，如《左传》中所谓“肉食者”都是指社会上层人士。而普通人则以素食为主，在富裕的年岁，年纪大的人才能吃上肉食。《孟子·梁惠王上》称：“鸡豚狗彘之畜，无失其时，七十者可以食肉矣。”

3. 秦汉时期

秦汉时期最大的变化是麦子在北方的种植更加普遍，这主要归功于水利技术的提高和磨的发明改进。麦子分为大麦和小麦，小麦的种皮硬不适合粒食，粉很黏，适合磨粉制作面食。大麦的种皮较软，粉不黏，适合粒食，可直接煮食，所以早期栽种的主要是大麦。在春秋末期、战国初期，磨已经发明。有学者认为当时可能只有水磨，磨出来的是浆类食物，还不能生产干粉。磨经过改进和发展后，小麦的可食性增强。因此，品质更佳、耐寒性更好的小麦代替了产量低、抗旱性差的大麦，在中原地区被广泛推广。南方地区延续了传统，一直是以稻子的种植为主。而且，这一时期稻子在北方一些适合种植的地区也开始推广。

大豆可能是当时百姓日常的主要食物。《汉书·货殖传》：“富者木土被文锦，犬马余肉粟，而贫者裋褐不完，含菽饮水。”这说的是富裕的阶层吃肉吃粟，但贫民只能吃豆子。大豆易成活、产量高。

秦汉时期肉类主要有两种来源，一种是草原地区的畜牧业，另一种是农耕地区的家畜养殖。秦汉时期，在与匈奴的战争中，中原地区往往能够获得大量的牲畜，以牛、马、羊等为主。在农耕地区，人们也会养殖牲畜，以猪、羊、鸡为主，间或有牛、马，但以牛、马为食物的较少。另外，还有鸭、鹅等家禽的养殖。从春秋末期开始，人工养殖水产品的现象产生，到了秦汉时期进一步发展。随着秦汉时期疆域的扩大，一些外来物种被引入中国，如葡萄、荞麦、苜蓿、豌豆等，丰富了中国人的餐桌，烹饪方法除传统的蒸、煮之外也出现了烤、炙等。随着发酵技术的发展，面食也不仅仅限于死面的做法，发酵之后制成的面食也开始出现。

4. 魏晋南北朝时期

魏晋南北朝时期是中华民族形成的重要时期，各民族之间的交流空前深入，在麦、稻种植进一步推广的同时，一些新的食品和烹饪方法被引入。

我们现在常见的黄瓜、茄子和胡椒就是这个时期被引入的。饮食文化专家李家文先生认为："黄瓜在中国有两种明显的生态型。南方型黄瓜直接由东南亚传入，现主要分布于华南地区，仍保持着要求温暖湿润气候的特性，而且一般为短日性；短日照有利于雄花分化，果实粗短，无明显的棱刺。北方型黄瓜约于2 000年前经由中亚细亚传入，经过长期在华北地区栽培，生态特性变异很大，它能适应北方的大陆性气候，耐变化剧烈的温度和干燥，除早熟品种外一般为长日性，能在长日照下分化雌花，而且果实细长，有明显的棱刺。"

茄子原产于印度和东南亚一带，在这一时期随着佛教被引入中国，并逐渐成为人们餐桌上一种常见的食物。茄子最早是在南方流行。

5. 隋唐五代时期

隋唐五代是对外交流最为频繁的时期之一，尤其是唐代国力强盛，对外交流的范围突破了西域各国，形成了一个辐射四周的"汉文化圈"。因此，在饮食结构上呈现了百花齐放的局面。

这一时期，最重要的变化体现在粮食作物上，稻米的地位跃升，成为与粟、麦等同等重要的粮食来源。稻米地位的提升，与自魏晋南北朝以来南方的开发和政治经济中心的南移有关。自魏晋南北朝以来，少数民族涌入北方地区，迫使中原人口向长江以南地区迁移，南方劳动力数量剧增，带动了南方地区的开发。安史之乱时，北方地区遭受战火破坏，而南方相对稳定，大量人口涌入淮河以南地区，进一步促进了南方地区的开发。南方逐渐成为中国的经济政治中心。

蔬菜种植方面，菠菜被引入中国，并得到了迅速推广。菠菜原产自今尼泊尔地区，因其耐寒易种植而被广泛栽种。唐朝人对菠菜已经有了比较深入的认识，唐代孟诜、张鼎撰写的《食疗本草》中记载："（菠菜）冷。微毒。利五藏，通肠胃热，解酒毒。服丹食人食之佳。北人食肉面即平，南人食鱼鳖水米即冷。不可多食，冷大小肠。久食令人脚弱不能行。发腰痛，不与蛆鱼同食，发霍乱吐泻。”

芥菜也是这一时期被引进的。此外，还有莴苣、扁桃、阿月浑子、树菠萝、无花果、西瓜、蚕豆、海枣（波斯枣）、海棕、枣椰、油橄榄以及香料丁香、孜然等。

6. 宋辽金元时期

宋辽金元时期，稻米成为全国第一大粮食作物，当时有谚语“苏湖熟，天下足”。这归因于自魏晋南北朝以来，人口的南迁。一方面，南迁的人群带来了先进的生产工艺和工具，使南方的生产力水平有了很大程度的提高；另一方面，人口的增加使很多荒地被开垦，加上南方有利的自然条件，南方自然成为“天下粮仓”。北方人的南迁，也促使麦子成为南方第二大农作物。

高粱是这一时期兴起的一种农作物，在金元时期已经成为北方重要的粮食来源。白菜在宋代成为主流蔬菜，主要在南方种植，经过长期的培育，到了宋代出现了包心大白菜，南北方均可种植，但以扬州地区的白菜口感为佳。

这一时期游牧民族进入中原，将大量农田改作牧场，对饮食上的直接影响就是羊肉和奶制品开始流行。不仅游牧民族，而且汉人也喜欢吃羊肉。北宋时期，上至宫廷下至民间都喜欢吃羊肉。

7. 明清时期

明清时期基本上延续了南方稻米、北方小麦的种植格局。在这一时期最大的变化是玉米、番薯、马铃薯等美洲农作物的引入。

玉米大概是在明朝嘉靖年间传入中国，到了清朝乾隆年间才普及全国。主要原因是清代人口数量激增，为了解决粮食问题，清政府大力提倡种植玉米。

番薯于明朝万历年间引入中国，明朝末年，福建和广东已成为著名的番薯产区。明朝以后，番薯种植在全国各地普及。

18 世纪末，马铃薯（土豆）被引入中国。一般情况下，与现在一样，大多数地区只是将马铃薯作为蔬菜食用。在灾荒时期及一些贫瘠的地方，马铃薯才被当作主食。

明清时期，被引入中国的还有辣椒、落花生、洋葱、番茄、西葫芦、南瓜、菠萝、向日葵、芒果、苦瓜等。

（二） 中国日常饮食结构的区域特征

民族的交流迁徙与地理环境对饮食文化的发展产生了极大的影响。从共时的角度出发，赵荣光先生将中华饮食文化分为以下几个区域："经过漫长历史过程的发展、整合的不断运动，中国域内大致形成了东北饮食文化区、京津饮食文化区、黄河中游饮食文化区、黄河下游饮食文化区、长江中游饮食文化区、长江下游饮食文化区、中北饮食文化区、西北饮食文化区、西南饮食文化区、东南饮食文化区、青藏高原饮食文化区、素食文化区。"[①] 对此，张景明先生做了略微的改动，删除了素食文化区，他认为："素食文化区不是一个地域性饮食文化区，应为散布于各个饮食文化区中的特殊饮食群体。"[②] 同时，他将"中北饮食文化区"改为"北方草原饮食文化区"，"中北饮食文化区，就是指以北方游牧民族为主体的饮食文化。在这一区域内的民族，以游牧和畜牧业为主要的生产生活方式。"[③]

1. 东北饮食文化区

根据赵荣光先生的划分，东北饮食文化区包括黑龙江、吉林、辽宁三省和内蒙古自治区靠近东北地区的部分区域。这一带天气寒冷，冬季时间长，作物生长时间短，但是土壤肥沃，水源、草原、山林资源丰富。生活的民族主要是汉族、蒙古族、满族及其他一些少数民族，其中汉族人来自山东、河南、河北等省。

东北地区饮食以面食为主，如饺子、馒头、包子等。因为肉类、鱼类资源

---

① 赵荣光、谢定源. 饮食文化概论［M］. 北京：中国轻工业出版社，2006：49.

② 张景明. 中国北方游牧民族饮食文化研究［D］. 北京：中央民族大学，2003：3.

③ 张景明. 中国北方游牧民族饮食文化研究［D］. 北京：中央民族大学，2003：3.

丰富，所以动物蛋白在饮食中的比重明显比中原地区等地方高。东北地区蔬菜类以白菜、萝卜、马铃薯等利于长期储存的蔬菜为主，大豆类制品如水豆腐、干豆腐、黄豆芽、绿豆芽等也比其他地方多。另外，东北地区气候寒冷，冷冻食品是当地的一大特色，如冻饺子、冻豆腐、冻水果等种类丰富。因为冬季时间长、天气寒冷，春夏季时间短，新鲜蔬菜比较缺乏，当地腌制类蔬菜、晾晒类食物很多，从而确保漫长的寒冷季节里的蔬菜供应，东北地区的泡菜也是全国知名。

冻水果

2. 京津饮食文化区

京津地区从宋元时期以来，一直是全国的政治中心。女真、蒙古、汉族、满族政权先后建都于此，因此全国各地的饮食都集聚于此。这里有牛肉、羊肉等北方游牧民族风味的食物，也有南方风味的菜品。尽管各类物品丰富，但京津地区地处北方，主食还是以面食为主，如馒头、包子、各类面点等。肉类以猪肉、牛肉、羊肉为主。春夏时节，各种时令蔬菜应有尽有；冬季严寒，多以白菜、土豆、萝卜等容易储存的菜品为主。

3. 北方草原饮食文化区

北方草原饮食文化区，是以内蒙古为中心，包括新疆、甘肃、宁夏、陕西、山西、河北、辽宁、吉林、黑龙江的部分地区。这一区域的少数民族主要有蒙古族、鄂温克族等，以游牧和畜牧业为主要的生产生活方式。

这一区域以肉食和奶酪为主。13世纪意大利传教士柏朗嘉宾在《柏朗嘉宾蒙古行纪》中曾有明确的描述："在牲畜方面，他们（蒙古族）都非常富有，因为他们拥有骆驼、黄牛、绵羊、山羊；至于牡马和牝马，据我看来，世界上的其他地区都不会拥有他们那样多的数量；他们不养猪和其他牲畜。""他们的食物是用一起可以吃的东西组成的。实际上，他们烹食狗、狼、狐狸和马匹的肉。……他们的食物中既没有面包，也没有蔬菜或可作蔬菜用的其他植物，没有任何这类食品，唯有肉类。"①

近年来该地区交通日益便利，在饮食上，除了肉类之外，米、面、蔬菜的比重也有所增加。在一些农耕区域，人们以粮食为主，肉类为辅。

4. 西北饮食文化区

馕

西北饮食文化区以新疆为中心，包括甘肃、青海、西藏的部分区域。人口以维吾尔族、哈萨克族、回族、蒙古族等少数民族为主。这一地区农业、畜牧业等协调发展，小麦、稻米是主要的粮食作物，羊肉是主要的肉类食品，还有葡萄、哈密瓜等闻名于世的瓜果。

饮食上，因为宗教信仰，牛羊肉和乳制品是当地的主要食物，还有馕、面条等面点，偶尔食用米饭。以前蔬菜的食用量比较少，近年来蔬菜的食用比重

---

① 耿昇，何高济. 柏朗嘉宾蒙古行纪；鲁布鲁克行纪［M］. 北京：中华书局，1985：30，41.

有所增加。

5. 黄河中游饮食文化区

黄河中游饮食文化区，包括陕西、山西、甘肃、河南、青海、宁夏的部分地区。这一区域是小麦的主要产区，因此饮食以面食为主，也以擅长制作面点而著称。如陕西的莜面、扯面，山西的刀削面，甘肃的拉面，河南的烩面等，都是全国闻名的面点。另外，还有陕西的锅盔、河南的烙饼等也是日常主食。

这一地区的肉类主要是猪肉，但一些回族聚居区域以牛羊肉为主。回族饮食对汉族饮食也有很大影响，如郑州、西安等地牛羊肉食品也非常受欢迎。蔬菜以当季的新鲜蔬菜为主，储藏、腌制菜品的数量较少。

6. 黄河下游饮食文化区

黄河下游饮食文化区，以山东为主，河南、安徽、江苏等邻近省份也包括在内。这一地区五谷杂粮都有出产，早期稻米较多，清朝中期以后玉米的种植量大增，替代稻米成为主要粮食作物。五谷杂粮磨成粉后制成的煎饼是普通百姓的主要食物，配以各类酱与蔬菜食用。该区域东部靠海地区，海鲜资源丰富，因此在日常饮食中占有一定比重。此外，日常饮食中也有小麦粉制成的馒头、包子、花卷、面条等面食。

7. 长江中游饮食文化区

长江中游饮食文化区指湖南、湖北、江西的部分地区。这一地区自然条件优越、水利资源丰富，是重要的粮食产区，很早就有“湖广熟，天下足”的说法。当地以稻米为主食，部分地区有少量的小麦生产，玉米、番薯的种植自清代以来比重日益增加。该地区有湘江、洞庭湖、东江湖等，因此水产品很多，湖南的剁椒鱼头、石锅鱼等都非常有名。肉类以猪肉为主，湖南的腊肉是全国闻名的美味。

8. 长江下游饮食文化区

长江下游饮食文化区指江苏、浙江、安徽、上海一带。这一地区是稻米的发源地，早在6 000~7 000年前的河姆渡文化遗址中就已经发现有大量的稻谷遗存，至今水稻仍是主要的农作物，米饭是日常饮食中的主食。长江下游河道密布，水产品丰富，鱼虾蟹之类，一年四季都有供应。沿海地区海鲜资源丰富，海产品在生活中占有一定比重。与北方相比，竹笋的食用比较广泛。肉类以猪肉、鸭肉等为主，蔬菜、果品等四季均有。

9. 东南饮食文化区

东南饮食文化区包括福建、广东、海南、广西以及台湾地区。这一地区地处亚热带，四季温差较小，水果、蔬菜资源丰富。该区域沿海，海产品也很多。日常主食以米饭为主。

10. 西南饮食文化区

西南饮食文化区指云南及贵州、四川的大部分地区。该地区以山地为主，水稻、小麦、玉米、高粱、番薯等均有种植，也都是人们日常生活中食物的来源。食物禁忌也比较少，猪、牛、羊、鸡、鸭、鹅等都有食用。一些少数民族还有食用蚂蚁、蚂蚱、蝌蚪等的习俗。

11. 青藏高原饮食文化区

青藏高原饮食文化区指青藏高原以及青海、四川、云南等邻近地区，以藏族为主体。该地区以农牧业为主，种植青稞、大麦、小麦等，后来又有玉米、水稻等，畜牧业以放养牛、羊为主。饮食上，牛羊肉、奶制品和面食都有，不同地区比重不同。风干、生冷的食物比较多，蔬菜之类的副食品较少。

二、烹饪方式

学会利用火是人类的一大进步，也推动了饮食制作方式的发展。中华饮食文化在长期的发展过程中形成了自己独特的烹饪方式，有煎、炒、烹、炸、蒸、煮等近百余种。在人们日常生活中烧烤、煮、蒸、炒是最基本的几种烹饪方式。

（一）烧烤类

烧烤应该是最原始的食物制作方式，最早可能是自然的原因，原始人吃到了烧熟或烤熟的食物。在学会使用火之后，他们才主动利用火来烧烤食物。

烧烤类的方法又可细分为烧、烤、炙、炮等。赵荣光认为：“原始人将大块的兽肉（以鹿科为主）挑或支架于篝火上，使食料直接与火接触，谓之烧；近火用热致熟则称为烤；炙，是将略小些的食料置于烧得温度很高的石块上致熟；炮，则是将不便于用烧、烤、炙等方法致熟的食料用泥等裹好投入火中或炭烬里，间接致熟。”[①] 在蒸煮等其他制作方式发明之后，烧烤依然是处理食物的重要方式，但是更多用于制作肉类制品。部分地区的面食也会使用烧烤的方式制作，例如新疆的馕、河南的烙饼、山东的煎饼。

烧烤

煨也是一种烧烤的方式，就是将食物埋藏于刚刚烧过的炭灰之中，利用炭灰的余热将食物烤熟。可以煨的食物有很多，以根茎类食物为主，这种方法一

① 赵荣光. 中华饮食文化［M］. 北京：中华书局，2012：126.

直沿用到近现代，玉米、马铃薯、番薯等都可以用煨的方式制作。

（二）煮类

煮大概萌芽于狩猎时代，最早的煮食将加热的石头投入盛有水的容器中，甚至是小水坑，利用石头的热量将食物煮熟。现代意义上的煮法要等到陶器的出现才真正兴起。赵荣光认为煮法是 7 000~8 000 年来中国最基本的烹饪方法，广泛用于菜肴和各类主食的烹制。而且用水煮食，可将米粒煮烂，使淀粉溶解在水里，营养得到保存，便于消化，味道也好。在长期的发展过程中，人们通过对火候和时间的调整，又衍生出炖、熬等不同煮法。

现代中国，煮食法遍布各个饮食文化区，粤菜中尤其以炖品出名，当地俗语有言："宁可食无菜，不可食无汤。"长期的炖品实践，形成了粤菜独特的风味。

（三）蒸类

蒸是利用水汽的热量烹饪食物的一种方式。中国人很早就发明了蒸食物的工具，称为"甑"，在距今 6 000~7 000 年的河姆渡文化遗址中，出土过底部有一个或多个穿孔的甑。

蒸可以用来蒸米饭，包括稻米、粟米等；也可以用来蒸面食，现代中国常见的馒头、包子、花卷、烧卖等都是蒸熟的。肉类、鱼类等也都可以用蒸法制作，如清蒸鱼、蒸羊羔、蒸鸭、蒸鸡等。湘菜中非常有名的腊味合蒸就是将腊肉、腊鸡、腊鱼等放在一起清蒸而成。蒸法也可以蒸蔬菜，如鲁南地区的香辣芹菜、粉蒸四季豆，江南的黄金莲藕等。

（四）炒类

炒法是现代中国最常见、最基本的烹饪方法，有清炒、煸炒、爆炒、小炒等不同的方式，但是炒法的出现却是相当晚的事情。从历史记载来看，先秦、秦汉时期食物制作的方式还是以烹煮为主，大约在魏晋南北朝时期炒法才开始普及。在现实生活中，食物制作往往并不是只用一种方式，而是综合利用多种方式。如鲁菜的锅塌黄鱼就是先将黄鱼炸好，然后放在锅里炖。东北菜“地三鲜”是先将土豆、茄子等在油锅里炸，然后再翻炒制成。

三、饮食惯制

（一）餐制

根据文献记载，先秦时期中国采用的是每日两餐制，只有早餐和晚餐。甲骨文中记载商人将一日分为八段，依次是：旦（或曰“明”）、大食、大采、中日、昃[1]、小食、小采、夕。这其中所谓“大食”“小食”就是指吃饭的时间，“大食”大概是7时至9时，“小食”大概是15时至17时。一日两餐制延续了很长时间，魏晋南北朝时似乎还有遗存。

赵荣光认为春秋时期已经出现了一日三餐的情况，明确地说一日三餐的是东汉时期的郑玄，在注释《论语·乡党》“不时，不食”一句时，郑玄解释道：“一日之中三时食，朝、夕、日中时。”可见，在汉代一日三餐比较普遍。现在，一日三餐也是中国各地最为主流的饮食习惯。

（二）饮食习惯

因为历史和地理环境，每个地区日常饮食中都有不同的特色和习惯。对一日三餐的具体安排也各有不同。

一般来说，北方地区早餐比较简单，主要是面点和相对易消化的一些汁水较多的食物。如北京地区早餐吃烧饼、馒头、包子等，也有油条、油饼、焦圈等油炸面食，再配豆腐脑、豆汁儿、馄饨、羊汤以及各种粥。天津地区饮食习

---

① 昃（zè）

惯基本与北京相近，比较有特色的是煎饼果子、面茶等。山东、河南也主要吃面食和粥类食物，山东比较有名的是煎饼；河南是胡辣汤，周口逍遥镇的胡辣汤最为出名。

南方早餐品种比较丰富，广东早茶最为出名。吃早茶是广东人的早餐习俗，不仅有各种茶，也有丰富的茶点。如今茶水逐渐退居其次，茶点却更加精致、丰富。最常见的茶点是包子，有各种品类，如叉烧包、水晶包、虾仁小笼包、蟹粉小笼包以及其他各类干蒸烧卖，各种酥饼，还有鸡粥、牛肉粥、鱼生粥、猪肠粉、虾仁粉、云吞、肠粉等。广西最有名的则是各种粉和油茶，长沙人早餐也酷爱吃粉，每天早上粉店都人满为患。

东部地区盛产稻米，早餐自然也少不了稻米，如上海的粢饭团，就是米饭中包裹咸菜和油条制成，也有放糖的。粢饭团与油条、大饼、豆浆一起被称为“四大金刚”。与北方地区明显不同的是，江浙地区早餐多吃面条，如舟山的海鲜面、宁波的面结面、苏州的焖肉面等。

西部地区以秦岭为界分为两部分，北部的陕西、山西地区口味与河南相似，有胡辣汤、羊杂汤、肉夹馍、各类粥等。南部的重庆、四川等地以麻辣鲜香为主。早餐有麻辣小面、抄手、米粉、米线、酸辣粉等。

一般来说，午餐和晚餐是人们一天中的正餐，是人们每天工作、学习所需要能量的重要来源。相对于早餐的简便而言，午餐与晚餐的菜品和分量都比较多。如今在一些大城市人们生活节奏比较快，一般午餐时间都比较短，虽然吃得多但是菜品并不精致，晚餐便成了一天中最重要的一餐。但是中小城市，尤其是北方大部分城市，中午一般有两个小时左右的休息时间，人们尽可以花心思做一顿丰盛午餐，晚餐则可以吃得简单随意些。

各地午餐、晚餐的种类丰富，包含肉类、蔬菜、主食、水果等。但是南北差异相对比较明显，北方主要是面食，以面条为主，包括卤面、拉面、烩面、刀削面等。在郑州，人们习惯在烩面馆或拉面馆叫上一碗面，有时候可以配上一两个凉菜。南方则会要一份米饭，然后，配上一两个炒菜，可荤可素。

口味方面，大致可以分为四种：北方多咸、南方和东方多甜、西北多酸、西南多辣。北方饮食中以咸鲜为主，日常饮食中调味料主要是盐，很少用糖。南方饮食中糖的使用比较普遍，有些人在炒蔬菜时也会加糖，不少北方人初到南方会明显感觉饮食上的不习惯。西北的酸主要是因为山西以醋闻名，有人说因为西北地区水土中钙含量丰富，多吃酸性食物可以避免得结石。能够吃辣的地区实际上不仅仅是四川、云南、贵州等西南地区，湖南人、江西人对辣的食物也非常钟爱。

四、饮食器具

饮食器具是饮食文化的核心内容之一，不同的工具影响着烹饪技术，比如炒的出现实际上得益于铁制锅具的发展。在漫长的历史进程中，中国饮食工具形成了包含陶、瓷、竹、木、金属等多种类型的综合体系。

陶制的饮食器具较早出现在人类的生活中，在旧石器时代和新石器时代，陶器在人类生活中发挥着重要的作用，不同色彩和用法的陶器也成了现代人区分不同历史时期文化特点的重要标志。在出土发掘的陶器中，盛食器、储藏器、盛水器、吹煮器是比较常见的饮食器，有壶、瓶、甑[①]、瓮[②]、罐、豆、樽、杯等多种器形。但旧石器和新石器时代并不只有陶器，在一些地区也曾出土石制或玉制的饮食器。

夏商周时期，陶器依然是主要的饮食器具，器形、纹饰变得更加精美。随着青铜工艺日益发展，青铜饮食器具也多有出土。如饮酒所用的爵，在夏商时期的遗址中都有发现。到了商代中期以后，青铜器皿的种类更加丰富，器形有鼎（烹调用具）、甑、鬲[③]、盘、豆等十几种。早期的青铜器制作较为粗糙、器

① 甑（zèng）
② 瓮（wèng）
③ 鬲（lì）

皿轻薄，商代后期至西周时期，器形变得厚重，造型也日益精美。玉器、漆器（包括竹器和木器）在春秋战国时期有了长足的发展，制作的器皿典雅、美观。

秦汉时期，青铜器逐渐退出了日常饮食领域，釉陶、竹木器等成为主流。目前出土的竹木器主要是盛、盒、樽、壶、耳杯、盘、勺、匕等。汉代铁制厨具开始使用，在广州南越王墓中就发现有各式铁厨刀四十余把，铁锅大概也是这一时期开始普遍使用的。

魏晋南北朝时期最值得注意的是瓷器的大量使用。瓷器比陶器更加坚固、精致，色彩更为丰富。在东汉时期，人们已经掌握了烧制瓷器的技术，但是到了魏晋时期才真正被大范围使用，到北宋时期真正成为饮食器具的主流。江苏省南京市长岗村东吴墓出土的青釉褐彩羽化升仙图盖壶，是我国目前发现最早的釉下彩绘瓷器，在瓷器发展史上被认为具有划时代的意义。

隋唐以后饮食器具的材质基本上已经定型，在造型方面的变化比较突出。随着中外交流的深入，波斯风格的器皿，具有契丹、女真、西夏等民族特色的器皿也出现在人们的餐桌上。另外，瓷器、玻璃器皿、金银等贵金属器皿在造型上也有丰富的变化。

现代中国饮食器具以铁器和瓷器为主，从类别上可以分为四大类：炊具、餐具、茶具、酒具。

炊具以金属材料为主，包括炒锅、煎锅、蒸锅、炖锅等。炒锅、煎锅主要是铁锅。铁锅一般不含其他化学物质，不会氧化。在炒菜、煮食过程中，铁锅不会有溶出物，不存在脱落问题，即使有铁物质溶出，对人体吸收也是有好处的。蒸锅除了铁锅外，还有用铝制成的。铁锅容易生锈，尤其是遇水时，铝锅轻薄、耐用、导热快、不生锈，所以蒸煮时比较方便。但是铝在高温状态下容易渗出，对身体有害。炖锅多是砂锅，砂锅通气性、吸附性较好，传热均匀，散热慢，适合炖煮食物。

餐具是吃饭的用具，如碗、筷、匙等。从材质来看，有金属、陶瓷、玻璃、纸制、塑料、竹木等，而以陶瓷为主。

茶具是喝茶用的器具，最基本的是茶壶、茶杯。另外，还有茶盘、茶盏、茶匙等辅助用品。

酒具是饮酒用的器具，常用的酒具是酒壶、酒杯。现代酒具主要是由瓷器、玻璃制成的，也有金属、玉石、竹木等制成者。历史上酒具有樽、壶、皿、鉴、斛、觥、瓮等不同的名称，实际上是根据造型而命名的。现代酒杯可分为三大类：一是白酒杯，器形小巧，一般容量在 30 毫升左右，多以陶瓷和玻璃制成；二是红酒杯，从西方引入，多是玻璃制成的高脚杯，容量在 300 毫升左右；三是啤酒杯，容量较大，一般在 500 毫升左右，多以玻璃制成。

第二章　节庆饮食

中国具有五千年的文明历史，在这么漫长的时间里，中国人创造了很多具有特色的岁时节日民俗，这些节日给人们的生活带来了许多欢乐，体现了人与自然之间、人与人之间和谐的亲密关系。其中最为丰富多彩的是节日饮食，每个节日都有其代表食物，每种食物里都包含古老的传统和丰富的寓意。

中国岁时节日食品大致可以分为以下三种：

第一种是用作祭祀的供品。供品在古代的宫廷、官府、宗族、家庭的特殊祭祀、庆典等仪式中占有重要的地位。在当代中国的多数地区，这种现象早已结束，只在少数偏远地区或某些特定场合还保留着一些象征性的活动。

第二种是供人们在节日食用的特定食物。中国的各类岁时节庆日从年初开始到年终，每个节日都有和习俗相对应的特定食品。如春节除夕，北方家家户户都吃饺子，寓含着亲人的团聚、阖家安康的意义和祝愿。端午节吃粽子则是把对爱国诗人屈原的怀念与对历史和乡土感情结合起来传承至今。中秋节吃月饼，圆圆的月饼象征天上的圆月，反映了人们期盼团聚的美好愿望。

第三种是节庆日和某些特定场合馈赠亲朋好友或其他对象的礼品。长期以来，中华文化以馈赠食品作为表达友好感情、建立亲密和睦人际关系的一种独特方式。例如中秋节亲朋好友间送月饼，端午节送粽子等都具有特定的民俗意义。春节期间人们在给亲朋拜年时也会带上节日礼物。

一、春节

中国“春节”这一名称是在1911年才出现的，但春节的历史则有3 000多年。农历正月初一，民间称“大年初一”，古代称“元旦”，民间过年从腊月二十三开始一直到正月十五元宵节，大约要持续一个月。农历年的最后一个月叫“腊月”，古代“腊祭”活动是为了答谢天地神和祖先。从腊月二十三开始，民间开始准备过年。腊月二十三或二十四，民间有“祭灶”的习俗。灶神是每户人家最重要的神，他监视着人间发生的所有事情，每年他都要上天向玉帝①汇报各户人家的情况。祭灶的食品中有一种叫“糖瓜”，据说是为了把灶神的嘴粘住，不让他在玉帝面前乱说话。过年期间，民间有在门口挂桃符、贴门神、贴春联、挂年画、放鞭炮烟花等习俗。

一年的最后一天叫“除夕”，除夕的晚上，家人一起吃一顿丰盛的“年夜饭”。这一餐的意义非比寻常，不仅是庆祝春节的到来，更是象征着团圆与和睦，预示来年全家生活幸福。无论大家相隔多远、工作多忙，在除夕这一天人们都会赶回父母的身边，和家人一起享用这顿团圆饭。年夜饭一般安排在年二十九（农历如果是小月）或年三十晚上。年夜饭又称“团年”或“合家欢”，因为这顿饭以后就要告别旧岁、迎来新岁了，所以又称“分岁”。在古代，人们认为年夜饭还有逐病、驱邪、健身的作用。年夜饭的每道菜肴都被人们赋予了极为美好的意义，如：火锅热气腾腾，温馨诱人，寓意为红红火火；“鱼”和“余”谐音，象征“吉庆有余”，也喻示“年年有余”；萝卜俗称菜头，祝愿有好彩头；虾、春卷等煎炸食物，预祝家运兴旺；甜点则是祝福以后的日子甜甜蜜蜜。

春节期间，通常都是亲朋好友一起聚餐，每顿饭都有酒有肉。在节日期间，可能人们就偏爱一些少油腻多清淡的菜肴。办春节宴席选用菜谱菜品要荤

---

① 玉帝：天上最大的神。

素搭配，冷菜、热菜、小吃兼顾。冷菜清淡不腻，可佐酒。热菜中既要有软熟、细嫩之菜，也要有香脆类的油炸食物。菜肴的色泽宜红、白、绿、黄相映，体现喜迎新春的心情；口味既有甜、咸，又有酸、辣，搭配合理，令人感到滋味无穷。

上菜顺序亦有讲究，前人曾有总结：“咸者宜先，淡者宜后；浓者宜先，薄者宜后；无汤者宜先，有汤者宜后。”一般应先上凉菜后上热菜，炒菜中要咸的先上，其鲜咸突出，能开胃，刺激食欲；油炸、甜食类可在中间或大菜之前上，因先吃油炸食品、甜食易使人产生饱腹感和油腻感；酸、淡汤菜应放在最后，既可醒酒，也可解腻；小吃安排在饭后，以其特色风味调整味觉，增加口感。现代人越来越注重饮食健康和营养搭配，春节期间也不再是暴食暴饮，饮酒过度，相反人们更加注意蔬菜和水果的摄入量，让吃进去的营养达到相对平衡，其代谢产物达到酸碱平衡，防止出现高血脂、体重骤增及其他不良后果。

春节宴席除了山珍海味等佳肴以外，美酒自然也是必不可少。阖家团聚的年夜饭，往往有老有小，每个人酒量不一，饮酒的习惯也不同。因此，可以同时准备白酒、红酒以及适合不饮酒人士喝的鲜榨果汁、玉米汁、牛奶等热饮。

以下是一些常见的具有特色的春节宴席经典菜肴：

凉菜：甜甜蜜蜜（蜜汁红枣）、春色满园（老醋苦菊）、百财兴旺（火腿白菜）

头菜：年年有余（清蒸石斑鱼）、鸿运当头（剁椒鱼头）、牛气冲天（铁板牛仔骨）、满堂笑哈哈（白灼基围虾）

热菜：团团圆圆（菜心羊肉丸）、红红火火（豆豉辣椒）、花开富贵（西兰花鲜鱿）、金玉满堂（松子玉米）

汤煲：雪梨香蕉汤、鲶鱼茄子煲、莴笋海米汤、雪花玉米羹

主食：饺子（流行于中国北方大部分地区，如韭菜鸡蛋素饺，寓意招财进宝）、年糕（流行于中国南方大部分地区）、步步高升（奶酪千层饼）、黄金果子、阖家幸福（扬州炒饭）

年夜饭菜肴本身也十分丰盛，鸡、鸭、鱼、肉俱全，鸡象征着吉利，鱼则代表年年有余。一些沿江或沿海地区，如上海、宁波等，年夜饭中还会出现很

多海鲜食品。除夕晚上中国人有“守岁”的风俗，“守岁”就是除夕晚上不睡觉，一家人一边吃饭、一边喝酒聊天或玩游戏，老人会给孩子压岁钱，并放在他们的枕头下面，希望孩子们来年身体健康，一切如意。

与春节有关的食物主要有腊八粥、年糕、饺子、元宵等。

1. 腊八粥

中国民间在“腊八节”当天有吃“腊八粥”的习俗。腊八节俗称“腊八”，即农历十二月初八，古人有祭祀祖先和神灵、祈求丰收吉祥的传统，一些地区有喝腊八粥的习俗。根据史料记载，我国喝腊八粥的历史始于宋代，距今已经有 1 000 多年的历史。每逢腊八这一天，无论是富人还是穷人，家家户户都要喝腊八粥。最早的腊八粥是用红小豆煮成，后来不同地域加上自己的地方特色，食材也逐渐丰富起来。腊八粥不仅是时令美食，更是养生佳品，尤其适合在寒冷的天气里保养脾胃。随着时代的发展，腊八粥的花样也越来越多，已经发展成为一种具有地方风味的小吃。

腊八粥

2. 年糕

春节期间中国人有吃年糕的习俗。春节吃年糕，寓意“一年比一年高”，希望大家在工作或生活上能够步步高升，越来越好。年糕的种类有很多：北方有白糕、黄米糕；江南有水磨年糕；台湾地区有红龟糕等。年糕主要用蒸熟的米粉经过揉、捣等工艺再加工而成。在江南，人们喜欢把糯米加水磨成米浆，蒸成条形或砖块状的水磨年糕。

据说年糕是从春秋战国时期吴国都城（现江苏省苏州市）传递到全国各地。宁波一带民间有“年糕年年高，今年更比去年好”的谚语。人们还用年糕印版压成“五福”“六宝”“金钱”“如意”等形状，象征“吉祥如意”“大吉大利”；有的则做成“玉兔”“白鹅”等小动物，达到真正意义上的内容与形式的完美结合。

3. 饺子

饺子源于古代的角子，饺子原名“娇耳”，是我国医圣张仲景发明的，距今已有 1 800 多年的历史。相传饺子是当时的张仲景为了让贫苦百姓免受冻疮之苦，用面皮包上一些祛寒的药材（羊肉、胡椒等）用来治病的，所以里面的馅主要是药材，起到药用效果。饺子是深受中国人喜爱的传统特色食品，是中国北方民间的主食和地方小吃，也是年节食品。饺子是财富的象征，在饺子里包藏银子或者元宝，象征着中国人对财富的渴望与憧憬。每年除夕一家老小热热闹闹围坐在一起包饺子，便有了齐心协力制造“元宝”的意思。晚上十二点的时候，人们就开始把包好的饺子投进锅里，这是新年的第一顿饭，寄托了人们对新年美好的愿望。

饺子多以冷水和面粉为料，将面和水和在一起，揉成大的粗面团，刀切或手摘成若干个小面团，将这些小面团用小擀面杖擀成中间略厚周边较薄的饺子皮，包裹馅心，捏成月牙形或角形，饺皮也可用烫面、油酥面或米粉制作；馅心可荤可素、可甜可咸；烹饪方法也可用煮、蒸、烙、煎、炸等；荤馅有三鲜、虾仁、蟹黄、海参、鱼肉、鸡肉、猪肉、牛肉、羊肉等，素馅分为什锦素馅、普通素馅等。饺子的特点是皮薄馅嫩、味道鲜美、形状独特、百食不厌。

饺子

二、元宵节

农历正月十五是中国传统元宵节，元宵节又称“灯节”。人们常常把过元宵节叫作“闹元宵”。元宵节那天晚上，到处张灯结彩，人们赏花灯、猜灯谜、看演出，不亦乐乎。

关于元宵节的起源，有很多种说法。一般认为元宵节与宗教活动有关。一种说法认为是汉代的时候，佛教从印度传入中国，当时的皇帝命令在正月十五这一天点灯，以表示对佛的尊敬。另一种说法认为正月十五是中国道教里一个神的生日，这一天人们点灯祭祀，希望能得到他的关心和爱护。还有人认为，元宵节与原始社会人们对火的崇拜有关。

元宵节的节日食物就是元宵，元宵本名汤圆，元宵节吃汤圆，表示“团圆如月”的意思。因为这是新年第一个月圆之夜，所以在这一天吃汤圆，就表达了希望一家团聚，一起共叙天伦的美好愿望。“元宵”作为食品，在我国也由来已久。宋代，民间流行一种元宵节吃的新奇食品。这种食品最早叫“浮元子”，后称“元宵”，生意人还美其名曰“元宝”。元宵在南方地区称“汤圆”，这种食品以白糖、玫瑰、芝麻、豆沙、黄桂、核桃仁、果仁、枣泥等为馅，用糯米粉包成圆形，可荤可素，风味各异。可汤煮、油炸、蒸食，有团圆美满之意。陕西的汤圆不是包的，而是在糯米粉中“滚”成的，或煮或油炸。

当然，元宵节也不仅仅只有元宵这一种节日食品，各地还有一些自己的特色食品，如东北人在元宵节爱吃冻果、冻鱼肉，广东人元宵节喜欢“偷”摘生菜，拌着糕饼，一起煮熟了吃，以求吉祥。

元宵

三、寒食节

我国古代在夏历冬至后 105 天、清明节前一二天还有一个重要的节日——寒食节。寒食节初为节时，禁烟火，只吃冷食。在后世的发展中，节日逐渐增加了祭扫、踏青、秋千、蹴鞠[①]、斗鸡等风俗活动，寒食节前后绵延2 000多年，曾被称为中国民间第一大祭日。寒食节是中国传统节日中唯一一个以饮食习俗来命名的节日。

寒食节传说是人们为了纪念春秋时期被大火烧死在山上的忠臣介子推而设立的。介子推，今山西介休人，晋国贤臣，后人尊为介子。介子推因不愿与趋炎附势的小人同朝为官，便携老母亲隐居山林。晋文公欲求却不得，放火烧山，原意是想逼介子推露面。结果，介子推抱着母亲被烧死在一棵大柳树下。为了纪念这位忠臣义士，于是晋文公下令：介子推死难之日不许生火做饭，要吃冷食，称为寒食节。但是因为寒食节与清明节邻近，加之寒食节期间各家各户不许生火，皆食冷食，食物容易变质对身体不利，所以唐朝以后寒食节与清明节合而为一，清明节渐渐成为祭祀祖先、追念先人的主流传统节日。

寒食节食品主要有以下几种：（1）寒食粥，寒食节期间，老百姓家家户户不许动烟火，于是节前将以大麦为主要材料的粥先提前煮好用于节日期间食用，故称为寒食粥；（2）润饼菜，又称嫩芽菜，是春卷的一种，发源于泉州，而后流行于台湾、福建等地区；（3）乌米饭，用糯米染乌饭树的汁煮成的饭，颜色乌青，也是寒食节主要食品之一；（4）欢喜团，四川成都一带有以炒米作团，用线穿之，或大或小，各色点染，名曰“欢喜团”，旧时在成都北门外至“欢喜庵”（欢喜指开心）一路摆卖。寒食节饮料有春酒、新茶、清泉甘水等数十种。不同地区寒食节食物也不同：晋南地区民间习惯吃凉粉、凉面、凉糕等。晋北地区习惯以炒奇（即将糕面或白面蒸熟后切成骰子般大小的方块，晒干后用土炒黄）作为寒食日的食品。在一些山区，这一天全家吃炒面（即将五谷杂粮炒熟，拌以各类干果脯，磨成面）。

寒食节，民俗要蒸“寒燕”庆祝，用面粉捏成大拇指一般大的飞燕、鸣禽、走兽、瓜果、花卉等，蒸熟后着色，插在酸枣树的针刺上面，装点室内，也作为礼品送人。

---

① “蹴鞠”指古人以脚踏、踢皮球的活动，类似现在的足球。

寒食粥

四、清明节

清明节是中国传统节日中唯一一个以节气命名的节日。清明节大约始于周代，距今已有 2 500 多年的历史。经过长期的发展，清明节具有极为丰富的内涵，各地都发展形成了不同习俗，而扫墓祭祖、踏青郊游是基本主题。中国人祭扫祖先的坟墓，怀念死去的亲人，体现了浓浓的亲情。同时，人们在这一天郊游踏青，感受春天的美好气息。

我国各地的清明食俗也丰富多彩。江南一带有吃青团的风俗习惯。青团是将雀麦草汁和糯米相互混合，然后包上豆沙、枣泥等馅料，用芦叶垫底，放到蒸笼内。蒸熟出笼的青团色泽鲜绿，香气扑鼻，是当地清明节最有特色的节令食品，也是祭祀祖先必备的食品。

青团

在浙江湖州，清明节家家做粽子，可作上坟的祭品，也可作踏青带的干粮。俗话说："清明粽子稳牢牢。"清明前后，螺蛳肥壮。农家有清明吃螺蛳的习惯，这一天用针挑出螺蛳肉烹食，叫"挑青"。吃后将螺蛳壳扔到房顶上，发出的响声据说可以吓跑老鼠。浙江桐乡河山镇有"清明大似年"的说法，清明当天晚上全家要在一起吃团圆饭，饭桌上少不了以下几个传统菜：炒螺蛳、糯米嵌藕、发芽豆、马兰头。吃藕是祝愿蚕宝宝吐的丝又长又好，吃发芽豆则是讨"发家"的口彩。

清明节各地还有食馓[1]子的风俗。"馓子"是一种油炸食品，口感香脆、外

① 馓（sǎn）

形精美，古代称“寒具”。现在流行的馓子有南北方差异：北方馓子大，以麦面为主。南方馓子外形精美，多以米面为主。在少数民族地区，馓子品种繁多、风味各异，尤其以维吾尔族、东乡族、纳西族、回族的馓子最为有名。

此外，各地清明佳节还有食鸡、蛋糕、清明粽、馍糍、清明粑、干粥等多种多样富有营养的地方风俗食品。如今因生活水平提高，多吃卤菜、盐茶蛋等。

五、端午节

端午节是中国第二大传统节日。端午节又称端阳节、五月节等，时间是农历五月初五，这是一个有着2 000多年历史的古老节日。“端”，是“初”的意思，“午”与“五”在古代是音义相通的两个字，所以“端午”其实就是“初五”的意思。

关于这个节日的由来有很多传说。不过流传最广的是纪念春秋时期伟大的诗人屈原。屈原是春秋时期楚国的大臣，因为楚国国君不听忠言，任用小人，最后导致国破家亡，屈原就怀抱大石跳入汨罗江殉国。他死后，楚国的老百姓非常悲痛，纷纷去汨罗江边悼念他。有的人划着船，在江上来回打捞他的身体；有的人拿出饭团、鸡蛋、粽子丢进江中让鱼虾吃，以免它们伤害屈原的身体；有的人将雄黄酒倒进江里，说是用药酒将蛟龙水兽弄晕，以免伤害屈原。从此之后，在每年的五月初五，就有了吃粽子、喝雄黄酒的风俗，以此来纪念爱国诗人屈原。

端午节除了吃粽子这一共同食俗外，各地还有其他丰富的食俗。例如：江汉平原一代吃黄鳝，甘肃吃面扇子，江西南昌吃茶蛋，河南、浙江农村吃大蒜蛋，吉林延边吃打糕，福建晋江吃煎堆，温州吃薄饼，四川会理吃炖药根等。

中国古人非常重视夏至和冬至这两个节气，因为这是季节交替的日子。夏至曾经是一个重要的节日，但是端午节产生以后，逐渐取代了夏至。夏至一般在公历六月中旬，过了夏至，夜就变长，白天缩短。这个季节经常发生洪水，害虫危害人们的健康，所以，古人认为五月阳气减少，阴气加重，各种“恶”，例如害虫、疾病都在这个月出现。民间认为五月是“恶月”，五月初五被古人看作最不吉利的日子。所以每逢端午人们要在窗上插艾叶、挂菖蒲。艾叶是一种芳香性化浊中药，有很强的驱瘟作用，古人称为“艾虎”。菖蒲也有祛病除邪作用，悬于屋檐下谓之“蒲剑”。过去许多人家这天还要用彩纸剪制“灭五毒”图案或吉祥葫芦帘花。女孩子用彩线和软帛制成精巧的五色粽子、老虎、蒜头等并将它们串在一起，悬在钗头或戴在小孩衣襟上。

与端午节有关的食物主要是粽子，另外在端午节这一天人们还要喝雄黄酒。

1. 粽子

粽子又称“角黍”“筒粽”，由粽叶包裹糯米及馅料蒸制而成。粽子用糯米做成，外面用竹叶包成角状。在漫长的历史演变过程中，粽子的“家族”也

粽子

不断壮大。形状上除了牛角粽，还有锥形粽、秤砣粽、菱角粽、包袱粽、枕粽，甚至在古代宫廷中还有楼阁亭子粽等。每逢端午节，中国百姓家家都要浸糯米、洗粽叶、包粽子，粽子花色品种更加多样。从馅料来看，北方多是包裹红枣、杂豆，还有包裹花生、松仁等干果的果粽，以及包裹蜜饯的蜜饯粽，江米粽、红豆枣泥、麦粽、核桃松仁粽等为北方代表。南方则有豆沙、鲜肉、八宝、火腿、蛋黄等多种馅料，火腿熏肉粽、干贝虾仁粽、闽台烧肉粽等，都是南方粽子的典型代表。

2. 雄黄酒

端午节这一天在古人心目中是毒日、恶日。因此，民间在端午节这一天喝雄黄酒也成为端午重要饮食习俗之一，特别是在长江流域。古语曾说："饮了雄黄酒，病魔都远走。"可见，喝雄黄酒的习俗已成为人们祈求平安、辟邪消灾的重要方式。

雄黄是一种传统的药材，也是一种矿物质，俗称"鸡冠石"，其主要成分是硫化砷，并含有汞，有毒。一般饮用的雄黄酒，只是在白酒或自酿的黄酒里加入微量的雄黄而制成，不可纯饮。雄黄酒具有杀菌、驱虫、解五毒的功效。"五毒"即蛇、蝎、蜈蚣、壁虎和蟾蜍，饮了雄黄酒便可杀"五毒"。中医还用雄黄来治疗皮肤病。在没有碘酒之类消毒剂的古代，用雄黄泡酒，可以祛毒解痒。所以，喝这种酒对人的身体有益，具有科学依据。大人们还用雄黄在儿童的额头点一下或画一个"王"字，既有杀菌的作用，也有辟邪的作用，寄托长辈对儿童的健康、平安成长的美好期望。

菜肴方面，虽不及春节、中秋那样讲究，但一般人家也要准备一些美味佳肴以应节令。端午的宴席以家宴为主，家宴设计可依家庭情况从简安排，以经济实惠为主。夏令时鲜蔬菜很多，可选用一些应时品种，比如瓠[①]子、蒜苗、黄瓜、鲜玉米等。在水产品方面，端午正是黄鱼、黄鳝上市时候，可依自己喜好选用。

---

① 瓠（hù）

点心方面，除了粽子外，还可做一些“五毒饼”（即带有五毒图案的酥皮点心）和“玫瑰饼”。端午时节，樱桃、桑葚、枇杷等已上市，这些以前都是必备的节令时鲜佳果。

饮料方面，除了准备啤酒和低度白酒外，节令专用酒有“菖蒲酒”。菖蒲酒以前是端午必备保健酒，由中药制成。现在此酒市场上经常可以见到，端午佳节不妨准备一些。万一买不到也可用桑葚酒代替。桑葚酒也是端午节令酒，具有“补五脏、聪耳目”之功效。

六、中秋节

中秋节在农历八月十五，又称“八月节”或“团圆节”，是流行于全国的传统文化节日。据说这天夜里月球距地球最近，月亮最大最圆最亮，所以从古至今都有宴饮赏月的习俗。

中国有 20 多个少数民族也过中秋节，而侗族把中秋节叫作“南瓜节”，仫佬[①]族称“后生节”，含义与汉族略有不同。

中秋节起源于中国古人对日月的崇拜。中秋节真正成为节日是在宋代。明清时期，中秋节逐渐成为和春节一样重要的传统节日。中秋节举行的活动有祭祀月亮和祈求丰收。在明代的北京，八月十五那天，人们要举行祭月的活动。人们用圆形的水果、饼供奉月神。北京民俗里的月神“兔儿爷”就是在那个时候产生的。

中秋节食俗以吃月饼为主要代表。与其他节日一样，在全国各地中秋节还有许多有特色的食物。例如：北京人吃醉蟹，福建人吃槟榔芋烧鸭，陕西西乡县人吃切成莲花状的西瓜，上海人以桂花蜜酒佐食，南京人吃桂花鸭，杭州人吃莼菜烩鲈鱼，四川人吃烟熏鸭子、麻饼、蜜饼，广东人吃芋头、糍粑等。

---

① 仫佬（mù lǎo）

中秋节美食首推月饼，圆圆的月饼象征“团圆”。月饼又称“胡饼”“宫饼”“小饼”“月团”“团圆饼”等，原本是古代中秋祭拜月神的供品，后来才慢慢演变为中秋节的必备食品。

“月饼”一词最早见于南宋文献。周密《武林旧事》卷六《蒸作从食》下罗列了许多“蒸作”的食品，其中有“荷叶饼”“芙蓉饼”“羊肉馒头”“菜饼”“月饼”等名目。吴自牧《梦粱录》卷十六《荤素从食店》下也列有“菊花饼、月饼、梅花饼”等名目。这两处记载都没有把“月饼”跟中秋节联系起来。考虑到《膳夫录》中专门记述中秋节的“节食”时提到“玩月羹”而不提月饼，同时也考虑到元代不见月饼的记载，那么宋人所说的“月饼”大概是一种月形普通食品，而非后世那种与中秋节不可分割的月饼。

真正明确提到中秋月饼的是明代文献。如田汝成《西湖游览志余》卷二《熙朝乐事》：“八月十五日谓之中秋，民间以月饼相遗，取团圆之义。”沈榜《宛署杂记·民风》“八月馈月饼”说：“士庶家俱以是月造面饼相遗，大小不等，呼为月饼。市肆至以果为馅，巧名异状，有一饼值数百钱者。”由此看来，月饼作为中秋节的节日食品被人们普遍认可是从明代开始的。所以，从明代中期开始，人们在中秋节互相赠送月饼已成为当时的重要礼仪。后来，月饼逐渐发展，形成了各地不同风格的月饼：京式月饼、广式月饼、苏式月饼、潮式月饼。月饼的种类越来越多，工艺也越来越讲究。咸、甜、荤、素各具风味；光面、花边，各有特色。例如：广式月饼皮薄、松软、香甜、馅美；苏式月饼松脆、香酥、甜咸适口；潮式月饼以糖冬瓜为馅，滋润、松脆。

中秋佳节也是阖家团圆的节日，家家户户都要备酒设宴欢庆佳节。中秋家宴的设计可依主人和客人情况，结合当地风俗而定。时令菜肴地区差异性很大。传统食品有卤馅、韭芽、烧卖、烧乳猪、南炉鸭、烧芋头、螃蟹、鳊花鱼等。果品中葡萄、槟子等节令佳果则不可少。白露节后，新栗上市，也可用其配制佳肴，如“五果炖鸡”。酒品可以选用桂花陈酒，晶莹明澈，有浓郁的桂花和葡萄酒的醇香。北京的桂花陈酒由于质量好、风味独特，曾多次被评为全国优质酒。

月饼

中秋宴会上，冷菜有姜丝拌肚丝、三色蛋糕、糖醋带鱼、开洋茭白、菜松等。热菜有葱炒肝花、珍珠虾球、蟹烧蹄筋、白玉饺子、板栗烧鸭、清蒸鳊花鱼。汤菜可以为五果炖鸡。主食可以为桂花炒饭，比较符合时令。

每逢中秋之夜，人们仰望着月中丹桂，闻着阵阵桂花香，喝着桂花蜜酒，欢聚于饭店或是家中庭院，其乐融融，是中秋佳节一种美好的享受。

七、重阳节

农历九月初九是中国传统的“重阳节”。重阳节，中国人有登高、插茱萸[①]、喝菊花酒的习俗。中国数字文化中，单数是阳数，九是阳数。九月初九，

① 茱萸（zhū yú）：一种植物。

两九重合，所以叫“重阳”。中国民间的传统观念认为，这是一个不好的日子，因此，那一天人们要“登高”“插茱萸”来消除这一天的厄运[①]。每年重阳节，有些地方的人就把家畜赶到外面。重阳节登高，传说是为了躲避灾难，实际上也是为了欣赏秋天的美丽景色；插茱萸也是为了躲避灾难，古人把茱萸插在头上，或者放在一个袋子里。

1989 年，中国政府正式确定九月初九为敬老节，从此登高望远和尊敬老人一起成为重阳节的两大传统。饮重阳酒和吃重阳糕都是重阳节的重要饮食习俗。

1. 菊花酒

菊花酒是一种适合在秋季饮用的养生饮品，古人认为菊花酒是“吉祥酒”。相传过去村里有一对夫妇，只育有一女，想再生一子，多年未得。重阳节前一天晚上，夫妇俩做了一个同样的梦：有人指点她（他），九月九日酿制一缸甜酒，七天后埋入地下，春节前取出享用，三个月后必有身孕。夫妇俩就按照梦里指示行事，第二年冬天果然得子，而且孩子聪颖异常，后高中举人。此后，重阳节酿制甜酒也就成了一种习俗。每到农历九月九日，天刚蒙蒙亮，主妇们便早早起来，淘米洗缸，男人劈柴挑水。到 9 点左右，男人将火烧旺，待水烧开，主妇便将糯米蒸熟，然后将酒药研碎。饭熟透后，主妇将饭倒置于洗净的团筛内散开，稍后将酒药撒在饭上、拌匀，在饭未凉之前盛于酒缸内、封好。用床单或棉絮捂上 3~5 天，有酒香味后将酒缸埋入地下。这样到春节前将酒缸取出时，缸里的酒便是色、香、味俱佳，而开缸时喷发而出的那股浓郁的酒香足以令人心旷神怡。

2. 重阳糕

有些地方重阳节这天，有戴菊花或吃重阳糕的习俗。重阳糕又称“花糕”“菊糕”。女儿要回娘家送重阳糕。人们也相互赠送重阳糕。工艺讲究的重阳糕要制成九层，像一座宝塔，糕上面还做两只小羊，以符合重阳（羊）之义。有

---

① 厄运（è yùn）：不好的运气。

的还在重阳糕上插一面红色小纸旗，并点蜡烛灯。这大概是用“点灯”“吃糕”代替“登高”。

据《西京杂记》记载，汉代已有九月九日吃莲饵的习俗，即最初的重阳糕。至宋代，吃重阳糕的风俗习惯开始普及。“糕”和“高”发音相同，因此人们用吃重阳糕的习俗来暗示重阳节“登高”的习俗。古代重阳节还是骑马练兵、讲武习射的节日，因此也有传说重阳糕就是从当时发给三军士兵的干粮演化而来的。

第三章　地方菜系

中国是一个历史悠久、地大物博且人口众多的国家，同时也是一个餐饮文化大国。长期以来，由于受到地理环境、资源物产、气候类型、文化传统等因素的影响，逐渐形成具有不同地域特色的菜肴种类，称为“菜系”。它以地域性的群体口味为主要特征，以独具一格的烹饪方法、调味方法、风味菜式、辐射区域、历史、文化为基本内涵。

一、菜系发展历程及形成条件

我国饮食文化历史悠久，可以追溯到先秦时期，由于地理、气候、物产、政治经济和风俗习惯等不同，在饮食风味上逐渐产生区别，当时就有了南北两大风味。秦汉以后，区域性地方风味食品的区别更加明显，南北各主要地方风味流派先后出现雏形。[①] 到了唐代，经济、文化高度发达，饮食文化也得到了长足发展，特别是唐代出现了高椅、大桌，使得人们在同一张桌子上吃饭成为可能。唐宋时期已经完全形成了南食和北食两大风味派别。和现代不同的是，当时北方人喜欢吃甜的，而南方人喜欢吃咸的。一直到南宋，由于历史原因，

① 杜莉. 中西烹饪历史比较［J］. 扬州大学烹饪学报，2002（3）：1－5.

大量北方汉人南渡到南方，并将甜食带到南方，这样甜食才慢慢成为南方人的主要口味，而这一时期中原北方[①]由于受到少数民族口味的影响，口味慢慢变得偏咸，从而形成了我们现在所熟知的“北咸南甜”。清代初期，川菜、鲁菜、粤菜、苏菜成为当时最有影响的地方菜，后称“四大菜系”。四大菜系分别属于三大文化流域，即黄河文化流域孕育了鲁菜、长江文化流域上游孕育了川菜、长江文化流域下游孕育了苏菜、珠江文化流域孕育了粤菜。[②] 清末时期，浙菜、闽菜、湘菜、徽菜四大地方新兴菜系分化形成，共同构成“八大菜系”，之后再增加北京菜、上海菜就有了“十大菜系”的说法。本章主要介绍中国的“八大菜系”以及别具特色的上海菜系。

菜系作为中国饮食烹饪的风味流派，在其形成过程中自然有一定的数量和品质的限制。综观中国的“八大菜系”，有以下几个形成条件。

（一） 最大限度利用当地原材料

古代中国，交通闭塞，取材范围很有限，当地人只能充分利用当地的原材料。因为原料特异，乡土气息浓郁，所以菜系风格往往别具一格，颇有吸引力。因此，不少菜系都非常注重原材料的开发。如北京的填鸭、山东的大蒜、四川的郫县豆瓣酱等，都是“我有你无”的标志性原料。

（二） 烹饪工艺的创新和独到之处

烹饪工艺是形成菜系的重要因素之一。中国向来以娴熟复杂的烹饪技艺著称，烹饪技法共有几十种，每种都有不同的特点。菜肴的色、香、味、形、质主要靠烹调技法来实现，一字之差，口味各异。不少菜系在风味上独树一帜，正是由于在炊具、火攻、形味和制法上有某些绝招，因此创造出一系列菜品。如鲁菜以制汤著称，徽菜以炖菜著称，川菜以炒煸闻名。由于技法不同，菜肴的口味和感觉截然不同。

---

① 当时的北方主要指现在的中原地带，如陕西、河南一带。

② 赵建民，金洪霞. 中国饮食文化概论［M］. 北京：中国轻工业出版社，2011：83.

（三） 众多名菜名点促进地方筵席发展

中国的“八大菜系”中每个菜系都拥有众多的名菜佳肴和面点小吃，这些美食加上各自的风俗习惯为地方筵席的形成奠定了基础，促进地方筵席民俗活动的形成与发展。由于筵席不仅是地方烹饪技艺的集中表现，同时也是地方民俗活动的凝聚点，因此是否有丰富多彩的地方筵席也是区分地方菜系的一个重要指标；同时，地方菜系的发展也促进了地方筵席的进一步发展。

（四） 具有持久的生命力和发展空间

中国地方菜系的形成不是一蹴而就的，而是在历史长河中经历了很长时间的洗礼和考验，是众多条件积累起来的结果。只有久经考验，不断积累，才能逐渐走向完善，达到成熟定型。在新时代下，交通发达、交流便利，不同菜系之间互相借鉴，在保持本身特点的前提下，不断创新发展，在发展中不断丰富、提升，以满足不同层次的需求，这正是一个菜系生命力旺盛的关键所在。

二、菜系形成的主要原因

菜系的形成与其悠久的历史及独特的烹饪特色有密切关系，同时也受地理环境、资源物产、文化传统、经济条件以及民族习俗等因素的共同影响。其中，地理环境、气候和物产等地域因素对菜系的形成与发展具有重要的决定作用。[①]

（一） 地理和气候因素

中国土地广阔、地形复杂，整体呈“西高东低”阶梯状分布，西部以高

---

① 赵荣光. 中华饮食文化［M］. 北京：中华书局，2012：29－30.

山、高原为主，中部以丘陵、盆地为主，东部以平原为主。气候也因为地理位置有所不同，东西递变为湿润、半湿润、半干旱、干旱。地理环境和气候的复杂性和多样性特征，形成了各地的食物原料和口味的差异。

1. 原材料的不同

中国有句俗话："靠山吃山、靠水吃水。"人们的食物大多是就地取材，特别是在古代，由于交通和生产力的落后，取材的范围更加有限，更容易形成相对固定的饮食文化圈。例如，东部沿海地区，因为靠近海域，盛产海产品，当地居民喜食鱼虾类食物，八大菜系为首的鲁菜就是以烹饪海鲜见长；而西北地区与海无缘，地形以高原为主，难以种植水稻，因此西北地区以面食为主。长江中下游地区地理环境优越，以平原为主，大小湖泊众多，当地的农民因地制宜种植生长习性喜高温多雨的水稻，是中国的"鱼米之乡"。

2. 口味的不同

由于地理环境和气候的差异，还形成了中国"东辣西酸，南甜北咸"的口味差异。吃辣椒主要和气候寒冷、潮湿有关，如我国的东部处于沿海地带，冬季阴冷潮湿，而四川地区虽不处于东部，但是由于地处盆地，一年四季少见阳光，同样潮湿寒冷，在这种湿润地区吃点辣有利于人们出汗，经常吃辣可以驱寒祛湿、养脾健胃，对健康极为有利（对当地人而言）。但是在东南沿海地区，如广州、福建一带，人们却很少吃辣，而喜食清淡的食物。这是因为这些地区长期以来天气炎热，人们容易上火生疮、生痰气瘀，所以需要食用清淡的食物帮助下火。

北方因为天气寒冷，蔬菜很难存放过冬，所以北方人想到用"腌制"的办法让蔬菜储存的时间更长一些，这样一来，北方大多数人也养成了吃咸味食物的习惯。而南方多雨，光热条件好，蔬菜生长更是一年几茬。南方人被糖类"包围"，自然也就养成了吃甜的习惯，像广东、浙江、云南等地居民也大多爱吃甜食。

另外，有些地方的饮食喜好和当地的水质有关。山西人爱吃醋，可谓"西酸”之首。原因是山西缺水，尤其缺乏稳定的优质水资源，许多地方是低洼盐碱滩，水土含碱量高。吃醋以“中和”碱性、缓释“水硬”的说法由来已久，形成了山西人喜酸的口味习惯。

（二） 历史文化因素

中华民族自古就有黄河流域文化圈、长江流域文化圈、珠江流域文化圈和辽河流域文化圈。不同流域其文化特色鲜明，因此逐渐形成了中原地带的雄壮辽阔，江南林园的秀丽优雅，华南沃土的华丽茂盛、西北地区的粗犷质朴，可谓是各有千秋。这些文化区域同时也是中国一些古城和繁华商贸中心的诞生地。如陕西西安、河南洛阳和开封、浙江杭州以及江苏南京都是中国著名的古都。这些古城在当时都是国家政治、经济中心，人口集中、商业发达，加上历代统治阶级对饮食的追求，逐渐形成了历史悠久、制作精良、难度高超的宫廷菜系，以鲁菜最为有名。江南一带环境优美、文人聚集，因此对菜系的色香味和就餐环境都有很高的要求，逐渐形成了精巧雅致的苏菜。西南的质朴之风则确定了川菜的朴实无华，川菜体现了“天府之国”的风貌。广州是历史悠久的通商口岸城市，吸收借鉴了外来的各种烹饪原料和烹饪技艺，使粤菜日渐完善，逐渐形成了兼容并包的特点。

（三） 宗教信仰和民族习惯

除上述原因外，宗教信仰和民族习惯的差异也会对地方菜系的形成产生一定的影响。中国是一个多民族大融合的国家，不同地区、不同民族的崇拜习性和信仰也影响当地居民对食料的选择和食用方法；同时，宗教信仰也会影响当地居民对食材的选用。中国自古就有把饮食生活转移到信仰生活中去的习俗，如佛教传入中国后，吃素食的饮食风俗在我国广泛流行，推动了蔬菜、瓜果类的栽培及豆类制品技术的发展，开创了我国饮食文化中素菜烹饪的一大流派。民族习俗和个人习性对饮食习俗也有重要影响。如满族的沙琪玛、维吾尔族的烤肉串、傣族的竹筒饭，均是不同民族饮食习俗的反映；而徽州人常年喜爱饮茶，所以徽菜一般油大，所谓“重油、色深、味浓”；西部地区的人们喜欢吃硬的食物，易造成消化不良，因而爱吃酸有助于消化，都表明了个人生活习性对菜系的影响。

（四） 生理和心理排外性

中华民族是一个重历史、重家族、重传统的民族，祖先留下来的东西世代传承，久而久之形成了一个地区的风俗。每个地区和民族对自己的饮食习惯都有着非常稳固的信念，轻易不会改变。这种心理因素的存在，使得各地区的饮食特征具有一定的稳定性和历史传承性。由于长期进食某类食物，人类的消化器官也发生了变化，这就形成了生理的排外性。北方人到了南方吃米饭，因为米饭不像馒头一样可以在胃中膨胀，所以有一种吃不饱的感觉。长期以植物性食品为主的人们，一连吃几顿肉，就会消化不良。因此，不同菜系都保持了各个地域的乡土特色。

“爱吃醋”这个词是怎么来的?

“醋”本来是中国人生活中非常重要的调味品之一，如山西人很喜欢吃醋是因为水质的原因。后来“吃醋”慢慢出现了引申义，表示“嫉妒”之义，表示男女相恋时有第三者介入，往往发生争风吃醋的现象。

那么“吃醋”这个词的引申义是怎么来的呢? 据说有这样一个典故，出自唐代宫廷。唐太宗李世民为了笼络人心，要为当朝宰相房玄龄纳妾，房玄龄之妻出于嫉妒，横加干涉，就是不同意。太宗无奈，只得令房夫人在喝毒酒和纳小妾之中选择其一。没想到房夫人确有几分刚烈，宁愿一死也不在皇帝面前低头。于是端起那杯“毒酒”一饮而尽。当房夫人含泪喝完后，才发现杯中不是毒酒，而是带有甜酸口味的浓醋。从此便把“嫉妒”和“吃醋”结合起来，“吃醋”便成了嫉妒的比喻义，流传至今。

三、菜系的主要分类

（一）鲁菜

鲁菜，是山东菜的简称，是中国菜系之首。鲁菜的形成与山东的地理环境、历史、文化、经济条件以及风俗习惯密不可分。从地理位置来看，山东是中国古文化发祥地之一，地处黄河下游，气候温和，胶东半岛突出于渤海和黄海之间。该地区山川纵横、河湖交错、沃野千里、物产丰富，为山东烹饪行业提供了取之不尽的物质资源，这些条件是鲁菜得以成为“菜系之首”极其重要的基础。从文化角度来看，山东又是“孔孟之乡”，文化底蕴丰富，儒家文化影响深远，使鲁菜在传播上占据了较大的优势，使得其影响力和传播广度是无与伦比的。从政治条件来看，山东所处的地理位置，与中原权力中心的位置最近，为它成为正统的宫廷御用菜系提供了便利的地理条件。

鲁菜的历史可以追溯到春秋战国时期，到了西周、秦汉，鲁国都城曲阜和齐国都城临淄都是相当繁华的城市，饮食行业盛极一时，名厨辈出；宋代以后鲁菜成为“北食”的代表；明、清两代，鲁菜已成为宫廷御膳菜，是八大菜系之首，同时也是技法最丰富、难度最大、最见功力的菜系之一。

鲁菜以清香、鲜嫩、味醇而著名，十分讲究清汤和奶汤的调制，清汤色清而鲜，奶汤色白而醇。一般认为，鲁菜由齐鲁、胶东、孔府三种菜系组成。齐鲁菜以香、嫩著称，尤其重视制汤，用高汤调制是齐鲁菜的一大特色；烟台福山为胶东菜发源地，以烹制各种海鲜而驰名，口味清淡，讲究花色；孔府风味以曲阜菜为代表，流行于山东西南部和河南，有“食不厌精，脍不厌细”的特色，用料很广，品种丰富。鲁菜常用的烹调方法有三十种，尤其擅长爆、扒。爆法讲究大火快烧，扒法是鲁菜独创。

鲁菜的代表菜主要有：德州扒鸡、糖醋鲤鱼、葱烧海参、九转大肠等。

德州扒鸡

德州扒鸡

特点：五香脱骨、肉嫩味纯、味透骨髓。造型上鸡的两腿盘起，爪入鸡膛，双翅经脖颈由嘴中交叉而出，整鸡呈卧状，色泽金黄。

糖醋鲤鱼

糖醋鲤鱼

特点：鱼尾翘起，色如琥珀，外焦里嫩。鲤鱼跳龙门的造型，酸甜的口味，表达了金榜题名、新婚花烛等人生大喜的欢欣之情。

葱烧海参

特点：海参新鲜、柔软、香滑，葱香浓郁，富含胶原蛋白却不含胆固醇，是不可多得的健康营养佳品。

九转大肠

特点：色泽红润，质地软嫩，兼有酸、甜、苦、辣、咸五味，是中餐中罕见的一道五味俱全的菜肴。

（二） 徽菜

徽菜是“安徽菜”的简称，是中国历史上典型的“因商而彰”的菜肴体系。“徽菜”因徽州商人的崛起而兴盛，又因徽商的没落而衰弱。徽州商帮的兴盛对饮食的讲究推动了家乡饮食业的发展，不仅使得徽菜的地位提高，成为宴请应酬的必备，也促使徽菜馆遍布全国各地，具有广泛的影响。徽菜的形成

葱烧海参

九转大肠

与古徽州独特的地理环境、文化内涵和饮食习惯密切相关。徽州风景优美、沟壑纵横、气候宜人的自然环境，为徽菜提供了取之不尽、用之不竭的徽菜原料。得天独厚的条件成为徽菜发展的有力物质保障，同时，徽州名目繁多的风俗礼仪、时节活动也有力地促进了徽菜的形成和发展。

徽菜的特点是就地取材，以鲜制胜，讲究火攻，重油重色，味道醇厚，保持原汁原味。徽州盛产河鲜家禽，就地取材使菜肴地方特色突出并保证鲜活。注重天然、讲究食补，这也是徽菜的两大特色。

徽菜的代表菜主要有：臭鳜鱼、清蒸石鸡、问政山笋。

臭鳜鱼

特点：闻起来臭、吃起来香，肉质鲜嫩、醇滑爽口，最大限度地保持了鳜鱼的本味原汁。

臭鳜鱼

清蒸石鸡

特点：汤清香郁，鸡肉细嫩柔滑、原味鲜醇。

问政山笋

特点：腊香扑鼻、笋香钻心、鲜香满席，热气腾腾，肥而不腻，咸中有甜，越嚼越香。

问政山笋

（三） 浙菜

浙菜是以杭州、宁波、温州等地的菜肴为代表发展而来的。与鲁菜悠久的历史和政治因素不同的是，浙菜主要以其所在地的文化色彩为特色。古语云："上有天堂，下有苏杭。"浙菜所在地正是这几个地方，风景优美，物产丰富，盛产鱼虾。浙菜的总体特点是清、香、脆、嫩、爽、鲜，注重原料的鲜、活、嫩。杭州菜以鱼、虾、禽、畜、时令蔬菜为主，讲究刀工、口味清鲜，突出本味。不少名菜来自民间，菜式小巧玲珑，菜肴的取名应景应情。宁波菜以烹制海鲜见长，讲究鲜嫩软滑，以炒、蒸、炖、腌制为主，注重大汤大水，保持原味。温州菜与宁波菜类似，也以海鲜为主，在烹饪方法上讲究"二轻一重"，即轻油、轻芡、重刀工，别具一格。

浙菜久负盛名的菜肴有西湖醋鱼、东坡肉、干炸响铃、西湖莼菜汤、龙井虾仁、宁波汤圆等。

西湖醋鱼

特点：西湖醋鱼选材精细，通常选用一斤半左右的草鱼。烹制时，火候要求严，仅用三四分钟烧得恰到好处。盛菜时，浇上一层糖醋汁。成菜后色泽红亮，吃起来鱼肉嫩美、带有蟹味，酸甜可口，别具特色。

西湖醋鱼

东坡肉

特点：菜品薄皮嫩肉，色泽红亮，味醇汁浓，酥烂而形不碎，香糯而不腻口。色、香、味俱全，深受人们喜爱。

干炸响铃

特点：选用轻薄的豆腐皮，卷入细末状肉馅后煎炸。成菜后色泽黄亮，形如马铃，松脆爽口。

东坡肉

干炸响铃

“西湖醋鱼”的民间故事

“西湖醋鱼”又称“叔嫂传珍”，其中有这样一段故事。相传在南宋年间，杭州当地有宋氏兄弟二人，都很有学问。但两人都不愿为官，隐居在山林，靠打鱼为生。哥哥已经娶妻，妻子很漂亮。当地有一个恶霸，见宋嫂年轻美貌，便害死了哥哥，想霸占宋嫂。宋家叔嫂祸从天降，悲痛欲绝。他们状告恶霸，不但没有告成，反而遭受毒打。回家后，宋嫂让宋弟赶紧逃到他乡，另谋生计。分别时，宋嫂特意用糖、醋做了一碗鱼，对宋弟说:“这菜有酸有甜，希望你有一天出人头地，不要忘记今天的苦难和辛酸。”后来，宋弟出走他乡，抗金卫国，当了大官，回到杭州，惩治了恶霸，但却不知道嫂子去了哪里。宋弟感到很愧疚，多年来一直寻找嫂嫂的下落。终于在一次吃饭中，席间有一道鱼，味道和他嫂子临别时做的那道鱼很相似，经询问才知道宋嫂隐姓埋名在饭店里当厨工，由此两人才最终团聚。

（四） 苏菜

苏菜，是江苏菜的简称。苏菜是长江中下游地区饮食风味的代表，历史悠久，文化积淀深厚，具有明显的江南饮食风味特色。苏菜由苏州、扬州、淮安、南京等地的菜肴构成。在整个苏菜体系中，淮扬菜占有很重要的地位，淮扬风味源自古城扬州和淮安，这里自古文人辈出，物产丰富，其中水产品尤为丰富，因此苏菜又称淮扬菜。

苏菜的特点是浓中带淡，鲜香酥烂，原汁原汤，浓而不腻，口味平和，咸中带甜。其烹调技艺以炖、焖、烧、煨、炒而著称。烹调时用料严谨，注重配色，讲究造型，菜式四季有别。苏州菜口味偏甜，配色和谐；扬州菜清淡适口，主料突出，刀工精细，醇厚入味；南京菜口味和醇，玲珑细巧，其中尤以鸭制的菜肴闻名。

苏菜久负盛名的菜肴有松鼠鳜鱼、清汤火方、扬州炒饭、蟹粉狮子头、盐水鸭等。

松鼠鳜鱼

特点：形如松鼠、外脆里嫩、色泽橘黄、酸甜适口。

松鼠鳜鱼

清汤火方

特点：汤清味醇、火腿酥香，其制汤方法能反映出扬州厨师高超的“吊汤”技艺。

清汤火方

扬州炒饭

特点：颗粒分明、粒粒松散，软硬有度，色彩调和，光泽饱满，配料多样，鲜嫩滑爽。

扬州炒饭

（五）川菜

川菜是四川菜的简称，川菜饮食文化是巴蜀文化的重要组成部分，它发源于古代的巴国和蜀国。川菜是目前最有特色且流传甚广的菜系。川菜历史悠久、源远流长，享誉中外，如今的川菜馆遍布中国各座城市的大街小巷，被誉为“百姓菜”。川菜的发展与四川独有的自然地理环境密不可分，四川位于长江中下游地区，四面环山，气候温湿，烹饪原料丰富多样。特别是因为气候湿润，所以通过吃辣椒可以排除体内湿气，保持身体健康。

一般来说，川菜以成都、重庆两个地方菜为代表，重视选料，讲究规格，层次分明。它取材广泛，善于用辣，离不开“三椒”（即辣椒、胡椒、花椒）和鲜姜，以辣、酸、麻为特点。川菜以“味”闻名，味型较多，富于变化，因此有“百菜百味”的特点，常见味道有家常味、咸鲜味、鱼香味、荔枝味、怪味等23种。烹饪方法擅长烤、烧、干煸、蒸。

川菜在国际上享有“食在中国，味在四川”的美誉，其代表菜有鱼香肉丝、宫保鸡丁、麻婆豆腐、夫妻肺片、回锅肉、东坡肘子等。

鱼香肉丝

特点：选料精细，成菜后色泽红润、有鱼香味，吃起来具有咸甜酸辣兼备的特点，肉丝质地柔滑软嫩。

鱼香肉丝

宫保鸡丁

特点：辣中有甜、甜中有辣，鸡肉的鲜嫩配合花生的香脆，入口鲜辣酥香、红而不辣、辣而不猛、肉质滑脆。

麻婆豆腐

特点：色泽光滑，豆腐白而嫩，味道麻而香。

宫保鸡丁

麻婆豆腐

“麻婆豆腐”的由来

清代同治年间，成都万福桥码头旁边有一家小馆子，老板娘脸上有麻子，人们都叫她陈麻婆。陈麻婆店里的主要客人是码头工人、脚夫。有一天，店快要关门的时候，又进来了一伙人，要求老板做点又下饭、又热、又便宜的菜。陈麻婆看店里没什么菜，只剩下几盘豆腐，一点牛肉末，现在去买菜肯定不行，已经太晚了。陈麻婆急中生智，把豆瓣剁细，加上豆豉，放油锅里炒香。加点汤，放入切成一指见方的豆腐块，再配上其他调料，加入炸酥脆的牛肉末，勾芡收汁，起锅以后再在豆腐上撒上一把花椒面，一盆麻、辣、烫、嫩、鲜的豆腐就上桌了。这伙人个个吃得鼻子冒汗，肚儿溜圆，口中大呼畅快。后来一传十、十传百，大家都知道陈麻婆做的豆腐又好吃、又下饭、又省钱，来的人越来越多，结果就成了陈家的招牌菜。因为这种豆腐又麻又辣，老板娘又叫陈麻婆，所以这道菜就叫陈麻婆豆腐。

（六）湘菜

湘菜，是湖南菜的简称。湘菜同样也是历史悠久，在汉代就已形成。湖南地处我国中南地区，长江中游，气候温暖，雨量充沛。与四川相似，因地理位置的原因，湖南气候温和湿润，人们多喜爱吃辣椒，用以提神去湿。

湘菜以湘江流域、洞庭湖区和湘西地区的菜肴为代表发展而来。湘菜历来重视原料互相搭配，滋味互相渗透，尤重酸辣。湘菜的共同风味代表是辣味菜和腊味菜，烹饪方法擅长腊、熏、煨、蒸、炖、炸等。著名的菜肴有剁椒鱼头、腊味合蒸、东安子鸡、冰糖湘莲等。

剁椒鱼头

特点：以鱼头的“味鲜”和剁椒的“辣”为一体。火辣辣的红剁椒，覆盖着白嫩嫩的鱼头肉，冒着热腾腾、清香四溢的香气。蒸制的方法，使鱼头的鲜香尽量保留在肉质中，剁椒的味道又恰到好处地渗入鱼肉当中，肉质细嫩、美味。

剁椒鱼头

腊味合蒸

特点：腊香浓重、咸甜适口、色泽红亮、柔韧不腻、稍带厚汁，且味道互补，各尽其妙。

腊味合蒸

东安子鸡

特点：造型美观、色泽鲜艳、肉质鲜嫩、酸辣爽口、肥而不腻、营养丰富。

（七）粤菜

粤菜，又称广东菜，虽然起步较晚，但是发展很快。不仅在中国有很多粤菜馆，国外的中国饭店多数以粤菜为主。粤菜的发展与广东的地理环境有很大关系，广东地处我国东南沿海一带，气候炎热潮湿，地形复杂，海岸群岛众多，海鲜品种多且奇。

总体而言，粤菜取料广泛。由于气候的原因，粤菜一般随着季节时令的变化而变化，夏秋偏清淡、春冬偏浓郁，讲究食补。因此，粤菜具有生猛、鲜淡、清美的特点。粤菜烹饪擅长煎、炸、炖、烩等方式，菜肴色彩浓而不腻，点心精巧、大菜华贵。代表菜有明炉烤乳猪、广州文昌鸡（鸡肉搭配鸡肝和火腿片）、白切鸡、白云猪手、蚝油牛肉等。

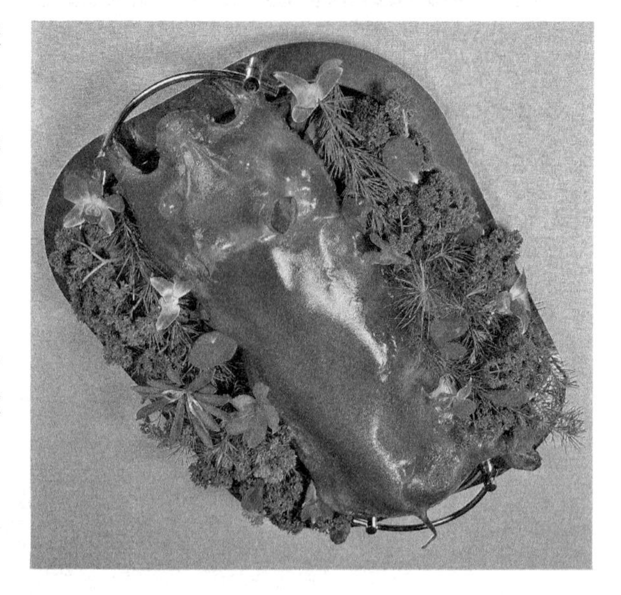

明炉烤乳猪

明炉烤乳猪

特点：乳猪色泽大红、油光明亮，皮脆酥香、肉嫩鲜美，风味独特。

广州文昌鸡

特点：文昌鸡造型美观，其肉质滑嫩、皮薄骨酥、香味甚浓、肥而不腻。

白切鸡

特点：色洁白带油黄，具有葱油香味，葱段打花镶边，食用时可以蘸芥末酱、酱油。

广州文昌鸡

白切鸡

（八）闽菜

闽菜是福建菜的简称。福建是我国著名的侨乡，华侨从国外引进一些新的食物品种和新奇的调味品，使其和我国传统的饮食相融合，不断丰富福建的饮食文化，从而逐渐形成带有开放特色的一种独特的菜系。闽菜是在福州、泉州、厦门等地菜肴的基础上发展而来的。闽菜的特点：一是擅长红糟调味；二是擅长做汤；三是注重刀工，有“片薄如纸、切丝如发、剞[①]花如荔”之美称。闽菜色调美观、滋味清鲜，主要的烹饪方法有炒、溜、煎、煨，尤以“糟”最具特色，闽菜中的红糟鱼、红糟肉、红糟鸡都是红糟菜肴体系中的代表。

闽菜的烹饪技艺，既继承了我国烹饪技艺的优良传统，又具有浓厚的地方特色。其风味特色是：清鲜，醇和，荤香，酸甜。代表性名菜有佛跳墙、龙身凤尾虾、醉糟鸡、太极明虾、荔枝肉等。

佛跳墙

佛跳墙

① 剞（jī）

特点："佛跳墙"是把几十种原料煨于一坛，既有共同的荤味，又保持各自的特色。吃起来软嫩柔润，浓郁荤香，荤而不腻；各种料互为渗透，味中有味。同时营养价值极高，具有补气养血、清肺润肠、防治虚寒等功效。

荔枝肉

特点：色泽带红、形似荔枝、质地脆嫩、酥香味佳、酸甜可口。

荔枝肉

醉糟鸡

特点：色泽淡红、骨酥脆、肉软嫩、味道醇香、食之不腻。

（九） 上海本帮菜

本帮菜是指上海本地的风味美食。上海本帮菜的一大特点是：浓油赤酱（油多味浓、糖重、色艳）。常用的烹调方法以红烧、煨为主，烹饪时糖用量大，口味咸中带甜，油而不腻。

本帮菜诞生之初主要服务于普通百姓，并不登大雅之堂。后来，本帮菜不断吸取外地菜特别是苏菜的长处，在二十世纪中叶形成了选料鲜活、口味适中的特点，许多本帮菜馆创出了看家菜，培育了一批本帮菜名厨，大大提高了本帮菜的品位。现在，受世界饮食潮流趋向于低糖、低脂、低钠的影响，本帮菜油、糖的投放量明显减少，以符合现代人饮食口味。因此，并不是所有的本帮菜都是浓油赤酱，也有随着季节时令变化而变化的清淡菜肴。另外，本帮菜烹调方法善于用糟，如糟鸡、糟猪爪、糟毛豆等，别具特色。

本帮菜中，荤菜中的特色菜有响油鳝糊、红烧甩水、油爆河虾、油酱毛蟹、红烧圈子、佛手肚膛、红烧回鱼、黄焖栗子鸡等，真正体现本帮菜“浓油赤酱”的特点。本帮菜中的蔬菜按季节不同有各种时令菜，如马兰头、荠菜、鸡毛菜、上海小油菜，这些蔬菜口味都非常清爽。上海传统小吃主要有小笼包、鲜肉月饼等。

红烧圈子

特点：色泽金黄、卤汁稠浓、香糯肥腴、嫩如面筋、咸中微甜。

红烧圈子

响油鳝糊

特点：鳝肉鲜美、香味浓郁、开胃健身。

响油鳝糊

红烧甩水

特点：以青鱼尾巴（即划水）红烧而成。菜品色泽红亮、卤汁稠浓、肥糯油润、肉滑鲜嫩。

红烧甩水

第四章　特色小吃

一、小吃形成的文化背景

中国土地广袤，民族众多，在中国境内广泛分布着众多具有地方特色以及民族特色的小吃。小吃作为中华美食的重要组成部分，在人们的餐桌上有着举足轻重的地位。最初，小吃多数由摊贩制作，以当地众多土特产作为原料，在街头销售以方便顾客，地方风味特别浓郁。

饮食文化的形成与变迁是地理环境差异和历史变迁长期作用的结果。特色小吃形成的背景也是如此。

（一）自然环境

自然地理环境制约作物类别从而影响食物特色，而且特殊的自然地理环境对饮食文化具有保护作用。其中，最典型的就是中国南北方的饮食文化差异。中国南北方的地理环境的差异，使各地区出产的作物有很大的不同。经过长时间的积累，形成了人们饮食上的差异，从而形成各具特色的地域性饮食习俗和传统，因而南北饮食习俗便有较为明显的不同。以黄河流域为代表的麦作地区饮食文化和以长江流域及其以南地区为代表的稻作地区饮食文化的差别尤为明显。因此，北方地区特色小吃多以面食为主，而南方地区特别是长江流域的特色小吃多以稻米为主要原料。如陕西地区的羊肉泡馍、荞面饸饹[1]、岐山面均以小麦为主要原料，而宁波汤团、上海排骨年糕、嘉兴粽子也均以糯米为主料。

（二） 气候条件

气候以及不同的土质影响人们饮食习惯和口味。如一年之中湿冷气候较重的四川、贵州、湖南当地居民爱吃辣，有祛寒除风湿的作用。山西黄土高原因含钙过多，因此当地居民爱吃酸，有利于消除体内的钙沉积，可以预防各种结石病。贵州许多少数民族喜欢酸食，这与长江中下游每年的梅雨季以及贵州山地气候所造成的“天无三日晴”有关系。如四川的担担面（四川的辣椒酱拌面）、湖南的口味虾都属于麻辣口味，贵州的酸汤粉和腌生（凉拌泡菜）则反映贵州饮食以酸为主的特点，而山西地区的饮食特点也可以用“无醋不欢”来概括。这既反映了气候、土壤对人们饮食习惯和口味的影响，同时也说明了饮食调理是人们适应自然环境的重要手段。

（三） 人文环境

人文环境如政治经济环境、宗教与民族环境都对地方特色小吃的形成有一定的影响。

1. 政治经济环境

先秦时期小吃就已出现，随着商业经济的发展进入快速发展时期并流行起来。小吃最早出现在一些重要的政治、经济、文化中心，例如繁华城市等。这些古代大都市政治局势安定、人口集中、商业繁荣，加之历代统治者讲究饮食，宫廷御膳、官府排筵、商贾逐味、文人雅集，这些都大大地刺激了当地烹饪技术的提高和发展。

2. 宗教与民族环境

我国现有宗教除道教是“土生土长”外，其他宗教都是历史上由国外传入

---

① 饸饹（hé le）

的，包括佛教、伊斯兰教、天主教和基督教。而对于饮食文化产生最大影响的宗教是佛教和伊斯兰教。我国共有 56 个民族，每个民族饮食习惯不同，因此各地小吃特色鲜明。不过即使在同一区域，也会因为各自民族的原始崇拜、图腾禁忌等原因使得小吃种类有所不同。比如在东北地区，满族和朝鲜族都是土生土长的少数民族，其生存环境、气候环境完全一致，但是饮食习惯却不同。比如，满族是不食狗肉的，因为满族人敬狗、爱狗，满族人多以打猎为生，狗所起的作用非同小可，久而久之就不吃狗肉了，满族神话故事中也有忠犬救主的故事。清太祖努尔哈赤将不吃狗肉列为一条不成文的法律。而生活在同一地区的朝鲜族则把狗肉作为很重要的饮食主料。由于宗教教规约束的饮食禁忌，也使得一些分布在相同地域内的不同民族表现出不同的饮食文化特点，而分布在不同地区的同一民族却表现出相同的饮食文化特点。如东北的满族和朝鲜族饮食文化不同，青海和甘肃一带的几个少数民族饮食文化也有差异，而分布在全国各地的回族在饮食文化上却具有共同性。

3. 文化融合因素

我国民族众多，在饮食文化发展过程中各民族饮食文化互相融合。如居住在中国新疆境内的哈萨克族的饮食主要以茶、肉（如纳仁[①]、熏马肠）、奶和面食为主。不过生活在城市里的哈萨克族除了上述的部分饮食外，还向周围的兄弟民族学会了做各种风味的饭菜，饮食结构趋于多元化。同时，反过来又影响了农牧区的哈萨克人，丰富了他们的餐桌。比如维吾尔族的米肠，乌孜别克族的羊肉、马肉、牛肉抓饭，回族的粉汤等都成为哈萨克人的家常便饭以及日常小吃。

二、小吃的形成过程

中国小吃是随饮食业繁荣流行开来的。在先秦《周礼》《招魂》中已有小

① 纳仁是新疆牧区的一种佳肴，具有明显的牧区特色。这种佳肴又称为手抓肉或手抓羊肉面。

吃名出现。小吃一般不作为正餐，它虽然被摆上筵席，随意作“点心”，但是也可以携之而食。不过这时的小吃享用的范围还极其有限，远不是市井或大众食品。汉代饮食行业兴起，“熟食遍列，肴旅城市”。市肆中售卖的小吃就有“豆饧”（甜豆浆）、“羊淹鸡寒”（酱肉熟鸡）之类。《史记·增殖列传》还记录有靠卖豆浆之类小吃发大财、富比公侯的人。

汉末发酵技术用于面食制作，面食的推广产生了蒸饼、胡饼（表面撒满芝麻的烤饼）、馒头、饺子等众多面食品种。这些面食集主副食于一身，价格不高，携带方便，市肆有售，因而百姓乐于接受。唐宋时期，小吃品种不断增多，并为不同层次人们享用消费。“花糕”由宫廷流传到市肆，也有专卖，应时性节令食品满足市民对季节性小吃的需求。既有胡饼店、蒸饼店、馄饨店、包子专卖店，也有食贩摊担，深入大街小巷，《水浒传》中武大郎炊饼担就属此类。此外，早市、夜宵、儿童零食等形式小吃出现。曹发发肉饼、太岁馒头等名牌小吃均已推出。到了明清时期，小吃则有了进一步的发展。如今随着经济发展，人们生活水平提高，吃在日常生活中占的比重越来越小，再加上工作节奏紧张，人们开始厌弃正餐而青睐小吃。如今，在一些城市还出现了小吃集中的民俗文化集散地，如北京的西四、大栅栏、天桥和王府井一带，天津的南市食品街，上海的城隍庙，苏州的玄妙观，无锡的崇安寺，南京的夫子庙都是几百年来闻名遐迩的“小吃群”，带有明显的市民饮食文化特色，成为独具魅力的重要旅游资源。

三、小吃的分类

（一） 按地域分类

中国作为国土面积世界第三的国家，地域广阔，这也决定各个地方的饮食文化差异极大，使我国饮食文化资源丰富。各地不仅有特色鲜明的名菜，也有具有地方特色的小吃。这些小吃不仅代表了当地的饮食文化，更多的是反映了当地的民俗文化和风土人情。从地域角度对中国地方小吃进行分类，可分为以下八个重点区域。

1. 京津风味

由于京津地区一带的特殊历史地位以及多民族饮食文化在此地交流融合，京津小吃在中国有着极其特殊的地位。

（1）北京小吃

由于北京特殊的历史地位，城市历史悠久，多民族的智慧积累，传统节日及宗教的影响，而且小吃取材极为广泛，加工制作技术多种多样和种类繁多的烹调方法，据不完全统计，历史上北京小吃的品种曾多达 3 000 种。如今，王府井小吃街、前门大街、南锣鼓巷都成为前往北京旅游的游客必去的地方。北京最负盛名的小吃有驴打滚（黄豆面饼）、豆汁和豌豆黄。

驴打滚

驴打滚是北方地区传统小吃之一，成品黄、白、红三色分明，煞是好看。因其最后制作工序中撒上的黄豆面，犹如老北京郊外野驴撒欢打滚时扬起的阵阵黄土，因此而得名“驴打滚”。

驴打滚

"驴打滚"的原料有大黄米面、黄豆面、澄沙、白糖、香油、桂花、青红丝和瓜仁。它的制作分为制坯、和馅和成型三道工序。做好的"驴打滚"外层粘满豆面，呈金黄色，豆香馅甜，入口绵软，别具风味。豆馅入口即化，香甜入心，黄豆面入嘴后可以不嚼，细细品。作为老少皆宜的传统风味小吃，"驴打滚"是你不想错过的一道美食。

豆汁

豆汁是老北京独特的传统小吃，最初是清宫御膳中的一种饮料，据文字记载距今已有300年的历史。

豆汁是以绿豆为原料，将淀粉滤出制作粉条等食品后，剩余残渣进行发酵产生的，虽味道特别，但具有养胃、解毒、清火的保健功效。夏天可消渴解暑，冬季能清热温阳，一年四季喝它，有益于开胃健脾、去毒除燥。老北京人喝豆汁儿有讲究，得配上焦圈、咸菜丝。这种特色小吃也形成了老北京独特的饮食文化。

豌豆黄

豌豆黄是北京春夏季节一种应时佳品。原为民间小吃，后传入宫廷。清代宫廷的豌豆黄，用上等白豌豆为原料，做出成品色泽浅黄、细腻、纯净，入口即化，味道香甜，清凉爽口。因慈禧喜食而出名。其制法是，将豌豆磨碎、去皮、洗净、煮烂、糖炒、凝结、切块而成。传统做法还要嵌以红枣肉。以仿膳饭庄所制最有名。

（2）天津小吃

天津得天独厚的地理位置，汇集了八方来客，也带来了八方美食，构成了天津特色的津门美食。提起天津的小吃，既有独特的色、香、味，又物美价廉，尤为天津人所乐道。天津的风味小吃多种多样，以南市、河北大胡同（今红桥大胡同）及鸟市等地最为著名，种类超过百余种。天津具有代表性的小吃有狗不理包子、耳朵眼炸糕和桂发祥十八街麻花。

狗不理包子

狗不理包子起源于清朝道光年间，传说是由居住在今天津市武清区的一位名叫狗子的蒸铺伙计首先制作而成。狗子的父亲 40 岁喜得贵子，中国人有给孩子“起贱名好养活”的传统，因此狗子的父亲就给自己儿子取名叫狗子（指狗）。狗子 14 岁开始在蒸铺做学徒，后来手艺精湛便独自开了一家包子铺，因为包子美味，所以生意十分兴隆，狗子每天都很忙碌。来吃饭的客人都说狗子太忙了，叫他都不理我了，于是狗不理（狗子不回应）的名声便传开。当时担任直隶总督的袁世凯吃了狗不理包子后大赞，并将包子献于慈禧太后。慈禧太后尝过后大赞：“山中走兽云中雁，陆地牛羊海底鲜，不及狗不理香矣，食之长寿也。”狗不理包子以每个包子都是 15 个褶为最大特点。

狗不理包子

耳朵眼炸糕

耳朵眼炸糕是天津地方名特食品之一，其制作精细，风味独特。这种炸糕外表和普通炸糕一样，扁圆形、金黄色。普通炸糕吃到嘴里是黏的，它却酥脆（外号“落地碎”），而且其内馅细嫩香甜，回味无穷，因而名扬国内外。此糕源起于清朝光绪年间，有个姓刘的串街走巷的小贩，和其外甥在耳朵眼胡同开了一家“刘记炸糕铺”。他为了把生意做好，招揽顾客，便选用黏黄米，精工磨细制成炸糕皮，而豆馅中红糖比一般的多，并将豆馅和红糖炒匀（普通炸糕没有这个工序），这种馅在夏天放半个月也不坏。此外，炸糕时是用温油，因此时间长糕皮酥脆，风味独特。

桂发祥十八街麻花

桂发祥十八街麻花是一道传统名点。天津市的百年老字号麻花店，与天津狗不理包子、耳朵眼炸糕并称“天津三绝”。桂发祥公司原创的什锦夹馅麻花，色泽金黄，外形像一根棍，夹着冰糖块，上面撒着青红丝和瓜条等小料，散发着香甜的桂花味。即使放上一个月，吃时仍然酥脆可口。

2. 东北风味

通常我们所说的东北地区指黑龙江、吉林、辽宁三省的行政区域。但是学术界所指的东北还包括内蒙古东北地区和河北北部地区。东北地区的历史源远流长，居住在东北地区古代的少数民族在与自然界长期搏斗以求生存的同时，创造了东北地区特有的饮食文化，它对促进我国南北饮食文化的融合以及世界东西方饮食文化的互动起到了积极的推动作用。东北饮食文化的形成与其自然环境以及民族融合有着密不可分的关系。东北具有代表性的小吃有烤冷面、粘豆包和水果冻食。

烤冷面

烤冷面是一道黑龙江常见的地方特色小吃，发源地为黑龙江省鸡西市密山市连珠山镇。油炸烤冷面制作方法简单，可以用鸡蛋、香肠等材料，主要由酱料作为调味剂。烤冷面兴起于 1999 年，据说是出现在密山市连珠山镇各学校门口，有碳烤、铁板烤和油炸三种做法，味道各有差异。目前主要流行铁板和油炸烤冷面。而铁板烤冷面可以添加鸡蛋、香肠、肉松、洋葱、香菜等，经过铁板的烘烤，涂上自己喜欢的酱料，松软可口、酱香扑鼻，吃起来很筋道，受大众喜爱。

粘豆包

东北粘豆包是我国北方满族人的一种民族传统食品，已有上千年的悠久历史，至今在我国广大北方地区仍然深受人们的喜爱。传统东北粘豆包多以黄米为主原料，采用自然发酵工艺制作而成，风味好、口感佳、营养丰富、抗饥耐寒，并逐渐演变成一种休闲和节日食品。

粘豆包

水果冻食

水果冻食的典型代表有冻梨、冻柿子、冰糖葫芦等。我国北方尤其是东北、内蒙古一带，冬季寒冷漫长，室外就是一个天然冷冻库，得天独厚的自然环境，为食品的保鲜与冷冻提供了良好的条件。冰天雪地造就了北方人的“冻食”嗜好，并形成了“冰雪饮食”习俗，且逐步发展成为冰雪饮食文化。冰雪饮食是冰雪生态环境下的产物，是北方人民对大自然充分利用的结果。

3. 西南风味

西南少数民族的小吃如同西南地区的“山地文化”一样带有浓厚的民族与地方色彩。我国有 55 个少数民族，西南地区有多个少数民族居住。云南、贵州、广西三地的少数民族人口约占三省区总人口的 31%。因此，西南地区的小吃有着独特的少数民族特色。西南地区具有代表性的小吃有担担面、过桥米线和酸汤鱼。

担担面

担担面是四川民间极为普遍且颇具风味的一种著名小吃，常由小贩挑担叫卖，由此得名（中文“担”意为“挑担”）。用面粉擀制成面条，煮熟后浇上炒制的猪肉末而成。以面条细薄，卤汁酥香，咸鲜微辣而著称。

过桥米线

过桥米线，以其制汤考究、吃法特别、别具风味而成为云南特有的风味小吃。过桥米线由汤、肉片、米线和佐料四部分组成。汤的表面有一层滚烫的

过桥米线

油，吃时要先把鸽蛋磕入碗内，接着把鱼片、鸡片、猪肉等各类肉片放入碗中搅动，好让生肉烫熟，然后放入各种配菜和调料，吃起来味道特别浓郁、鲜美。

酸汤鱼

酸汤鱼，是贵州、广西、湖南交界处的一道侗族名菜，与侗族相邻的苗族、水族、瑶族等少数民族也有相似菜肴，但其中以贵州侗族酸汤鱼最为有名。据考证，此菜肴源于黎平县雷洞镇一带。制作原料主要有鱼肉、酸汤、山仓子等香料。成菜后，略带酸味、幽香沁人、鲜嫩爽口开胃，是贵州“黔系”菜肴的代表作之一。这道菜通常先自制酸汤，之后将活鱼去掉内脏，入酸汤煮制。

4. 东南风味

东南地区风味小吃以广东小吃和福建小吃为主。其特点多为造型精细讲究，花式品种令人眼花缭乱，原料多以稻米和海鲜为主，这与其气候特点和地理位置都有密不可分的关系。东南地区具有代表性的小吃有顺德双皮奶、海蛎煎和七星鱼丸。

顺德双皮奶

双皮奶始创于清朝末年，是当时广东顺德的一位农民无意中调制出来的。顺德双皮奶制法考究，奶白而滑，香味浓郁，质感细腻，可搭配红豆、莲子、芒果、芝麻糊一起吃，甜甜蜜蜜的滋味让人难忘。

海蛎[①]煎

海蛎煎是一道常见的家常菜，起源于福建泉州。福建的做法是，在烧热的铁板上放几个新鲜肥硕的海蛎做快速的煎炒，勾芡再加上一些小白菜、豆芽菜，打一个鸡蛋。而作为台南一带的传统点心，是以加水后的番薯粉浆包裹蚵仔、猪肉、香菇等食材煎成饼状物。2018 年 9 月，海蛎煎被选为福建十大经典名菜。

七星鱼丸

七星鱼丸是一种包馅的鱼丸，源于清朝初年。鱼肉剁成茸状后加薯粉搅拌

---

① 蛎（lì）

均匀，以猪肉作馅制成球形丸子，在汤中煮熟后浮沉摇摆，似空中星斗，故名“七星鱼丸”，为福州名点。

5. 江南风味

江南风味小吃主要以各类点心为主。长江下游江浙沪一带所制作的米面食品，以江苏为中心，故称苏式点心，又称“江南点心”（长江以南地区的点心）。由于江南地区处在富庶的鱼米之乡，经济繁荣，物产丰富，饮食文化发达，为制作多种面点提供了得天独厚的条件。江南地区负有盛名的点心有蟹黄汤包、汤圆和粽子。

镇江蟹黄汤包

镇江蟹黄汤包，俗称蟹包，是镇江的传统名点，相传已有200多年的历史，不但在上海、南京负有盛名，而且驰名中外。这种汤包以蟹油、猪肉为主要原料，经过精心加工制成。它体积小、外形美，放在笼里像座钟，夹在筷上像灯笼。皮薄、汤多、馅足，食时佐以镇江香醋与姜丝，不但口味更美，而且能去腥解腻。

宁波汤团

汤圆是浙江宁波著名传统小吃之一，也是中国的代表小吃之一，历史悠久。传说，汤圆起源于宋朝。当时宁波盛行吃一种新奇食品，即用各种果饵做馅，外面用糯米粉搓成球，煮熟后，吃起来香甜可口。因为这种糯米球煮在锅里上浮下沉，所以它最早叫“浮元子”，后来有的地区把“浮元子”改称“元宵”。与北方人不同，宁波人在春节早晨都有合家聚坐共食汤圆的传统习俗。

嘉兴五芳斋粽子

五芳斋粽子品牌创立于1938年，至今已有80多年历史。五芳斋粽子号称“江南粽子大王”，以糯而不烂、肥而不腻、肉嫩味香、咸甜适中而著称。五芳斋粽子按传统工艺配方精制而成，选料十分讲究，肉粽选用上等白糯米、后腿瘦肉，甜粽则用上等赤豆“大红袍”，通过配料、调味、包扎、蒸煮等多道工序精制而成。嘉兴五芳斋粽子有肉粽、豆沙粽、蛋黄粽等几十个品种。

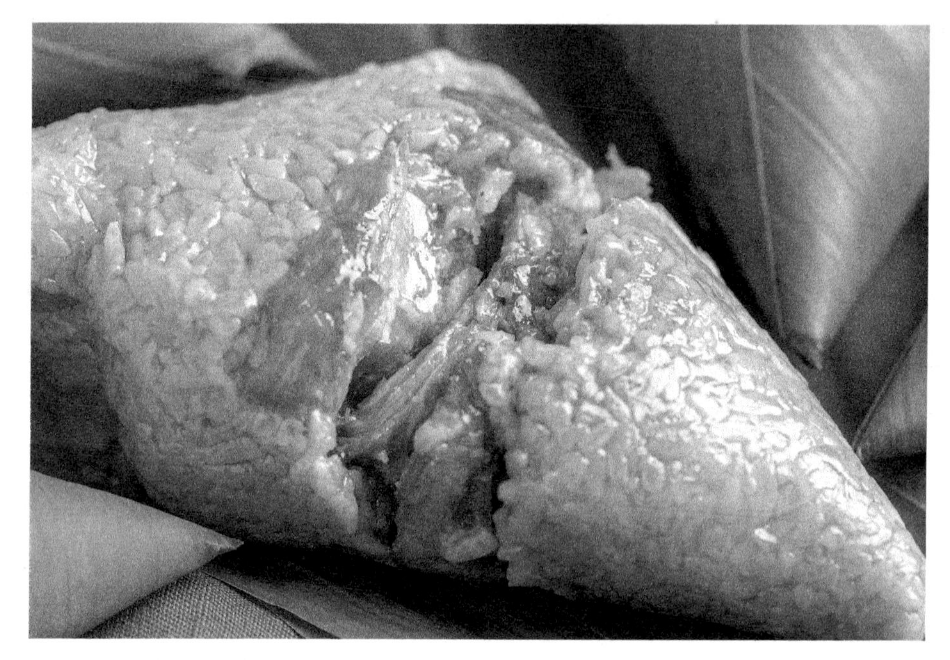

五芳斋粽子

6. 湖北、湖南风味

湖北和湖南小吃是我国中南地区风味的代表。当地人嗜辣，小吃也多以辣味为主，兼有咸、甜味，令人食欲大开。两湖地区具有代表性的小吃有热干面、臭豆腐和口味虾。

武汉热干面

武汉热干面与山西刀削面、两广伊府面、四川担担面、北方炸酱面并称为我国五大名面。面条先煮熟，然后过冷、过油，淋上用芝麻酱、香油、香醋、辣椒油等调料做成的酱汁，添加虾米、五香酱菜等配料，增加多种口味，吃时面条爽滑筋道、酱汁香浓味美，让人食欲大增。

臭豆腐

长沙人称臭豆腐为“臭干子”，黑乎乎的颜色，怪异的气味，完全不同于其他地方的臭豆腐。长沙臭豆腐闻起来臭，吃起来香，外酥里嫩，辣味浓厚，一经品尝常令人欲罢不能。

臭豆腐

口味虾

口味虾是湖南特色小吃，以小龙虾制成，口感麻辣鲜香，更是夏夜喝啤酒时必不可少的佐酒食物。夏天夜幕降临的时候，长沙的街头巷尾到处都能看到正在品尝口味虾的食客，尽管个个都被辣得嘴巴通红、眼泪汪汪、满头大汗，却依然兴趣不减。

7. 西北风味

西北风味以山西小吃和陕西小吃为代表。西北地区人们普遍爱吃面食，小吃也以面食居多，味道醇厚，有浓郁的乡土特色。西北地区具有代表性的小吃有羊肉泡馍、陕西凉皮和猫耳朵。

羊肉泡馍

羊肉泡馍是西安最有特色、最有名的食品。将白面烤饼掰碎泡进羊肉汤里食用，肉汤鲜美，馍也入味。羊肉泡馍烹制精细，料重味醇，肉烂汤浓，肥而不腻，营养丰富。因为它暖胃耐饥，所以为西安和西北地区人民所喜爱。西安的羊肉泡馍馆很多，其中以老字号“老孙家”“同盛祥”较有名气。

羊肉泡馍

陕西凉皮

陕西凉皮分为大米面皮和小米面皮两大类，因大米面皮最受欢迎，故又称米皮。一般吃时把面皮切成半厘米宽的条，加辅料黄瓜丝，调入盐、醋、酱油、芝麻酱、辣椒油等调拌即可，酸香可口，夏天吃尤为清爽解暑。

猫耳朵

山西小吃猫耳朵是把和好的面分成蚕豆大的小块，用拇指、食指捏着一转，便被卷成像猫耳朵一样的形状。在开水里煮熟它，捞起来配以韭菜、肉丝、虾仁、冬菇、火腿等大火炒熟，耳卷里吸存着汤汁，味道饱和，吃起来十分鲜美。

8. 上海风味

上海是长江经济带的龙头城市，是全球著名的金融中心，世界上人口规模和面积最大的都会区之一。上海的城市性质使得这里变成了各种各样小吃汇集的地方，集合了全国甚至全球的风味小吃又加以自己的特色。如《上海通史》第五卷《晚清社会》，其中的第八章“中西混杂的都市习俗”中有一部分描写晚清上海饮食业的发达与饮食新风尚。书中分析了开埠后上海“饮食从满足生理需求和简单娱乐为主转向重在以饮食为交友之道，追求舒适高档和多种口味”。书中指出，“造成这种状况的原因是受太平军战事影响，租界人口的增加和大量流动人口的存在，饮食需求扩大。全国各地移民的涌入带来了不同口味的需求，使上海饮食业从单一走向多元化。”由于上述历史原因，上海小吃也有这类特点，但依旧保持了自己的传统特色。上海具有特色的小吃有生煎、小笼包、排骨年糕和四大金刚。

生煎

生煎又称生煎包子、生煎馒头，是流行于上海、浙江、江苏及广东等地的一种传统小吃，广受国人喜爱，制作材料有面粉、芝麻、葱花、猪肉、肉皮冻等。生煎原为茶楼、老虎灶（开水店）兼营品种，馅心以鲜猪肉加肉皮冻为主。20 世纪 30 年代以后，上海饮食业出现了生煎的专营店，馅心的花色也增加了玉米、荠菜、虾仁等品种。现如今，上海的“小杨生煎”闻名全国。

生煎

南翔小笼

南翔小笼原名“南翔大肉馒头”“南翔大馒头”“古猗园小笼”，也称“南翔小笼包”“南翔小笼馒头”，是上海市嘉定区南翔镇的传统名点，以皮薄、肉嫩、汁多、味鲜、形美著称。南翔小笼的馅心是夹心腿肉制成的肉酱，仅撒少许姜末和肉皮冻、盐、酱油、糖和水调制而成；皮是用不发酵的精面粉制作而成的。2014 年 8 月，文化部第四批国家级非物质文化遗产代表性项目名录推荐项目名单公示结束，南翔小笼制作技艺成功入选全国传统面食制作技艺。

排骨年糕

排骨年糕是上海一种价格实惠、独具风味的传统小吃，已有 50 多年历史。大排佐以小而薄的年糕，经油氽[①]、烧煮而成，既有排骨的浓香，又有年糕的软糯酥脆，十分可口。排骨色泽金黄，表面酥脆，肉质鲜嫩。年糕入口糯而香，略带甜辣味，鲜嫩适口。

四大金刚

说起上海的早餐，最著名就是“四大金刚”，即大饼、油条、豆浆、粢饭团。

大饼一般有两种口味：甜和咸。甜大饼是用白砂糖做的馅，表面撒上白芝麻，香甜可口，外表焦香。咸大饼搭配葱花，制造出美味，这是很考验技术的一种做法。

油条不仅是在上海出名，而且在全国各地都有名，在上海多用大饼将油条包裹食用，或与豆浆和粢饭团搭配。

同北方主打甜味不同，上海的豆浆分为咸甜两种。甜豆浆是直接在豆浆中加入白糖。咸豆浆是搭配虾皮、油条、榨菜、生抽或者特调的酱汁，再滴上几滴麻油，即可食用。

粢饭团以糯米饭为主原料，裹进油条、榨菜、肉松等卷捏而成，味香而有嚼劲。以前粢饭团里加糖或黑芝麻等，后来加入榨菜、肉松。

---

① 氽（cuān）

（二） 按食材分类

地理环境因素会对不同地区居民的生产、生活产生重大影响，这种影响也体现在饮食习惯上。由于我国疆域辽阔，不同地区地形、气候差异较大，各地种植农作物种类和养殖畜牧业种类都有不同，因此根据食材分类，小吃也可分为多种类型。本节从稻米、麦类、肉类、海鲜四个方面对知名小吃进行简要介绍。

1. 稻米

现代考古研究表明，新石器时代，居住在中国东南沿海地区的祖先就开始栽培水稻以生存繁衍。以稻为基础的文化（稻文化）随着人口的迁徙，与产生于黄河流域的北方旱地文化（粟文化）交融，共同孕育了灿烂的中华文明。稻在中国的地位和稻文化是随历史的发展不断提高和发展的。在古代，粟、麦、稻是三大主要粮食作物。从新石器时代晚期到商周时期粟居首位，麦次之，稻最后。到秦汉时期以后，麦的地位上升，接近粟的地位，稻仍为末位。从三国到魏晋南北朝（公元 3 ~ 6 世纪），因为南方的开发加快，稻的比重上升，与粟、麦同等重要。唐宋时期（公元 7 ~ 13 世纪）及以后，稻便取代粟麦而居首位。

2. 麦类

中华民族有着以各种谷物淀粉为原料制作条形食品的悠久传统，据考古发掘，距今 4 000 年的中国青海喇家索面是迄今为止发现最古老的面条实物，是用小米和黍两种谷物制成。历史上，小麦和大麦最初是像稻米一样粒食的。小麦制粉在两汉时期有了长足进步，因此面食品种激增，馒头、饼、面条、包子、饺子等的初期形态竞相出现，面粉的发酵技术也随之发明。中国以各种麦类谷物（主要是小麦）作为主要食材来制作的小吃不胜枚举。

3. 肉类

悠悠远古，华夏先民“食草木之实、鸟兽之肉，饮其血，茹其毛”，长期过着生吞活嚼的原始生活。传说后来有“燧人氏”发明人工取火，教人用火熟食，“以化腥臊”，才开始了真正意义的“烹”，食物的品种和范围也就扩大了。中国肉食文化的发展同烹饪技术的提升以及火的使用有着极为紧密的关系。中国幅员辽阔，气候和地理环境多样且复杂，中国北方畜牧业发展较好，同时各民族饮食文化的相互融合也使中国的肉食文化有了快速的发展。以各种“肉”作为食材而制作的小吃品种非常丰富。如新疆的羊肉串、上海的南翔小笼包和排骨年糕都是以肉类为重要原料。

4. 海产品

中国大陆海岸线全长约 1.8 万千米，如果再将所有岛屿的海岸线相加之后，则有 3.2 万千米。如此长的海岸线，为中国饮食文化提供了多样的海产品选择。以海产品为主要食材而制作的小吃品种丰富。如福建厦门的沙茶面、七星鱼丸都是以海鲜为主要制作原料的小吃。

第五章　茶酒文化

一、饮茶

（一）茶的故乡

1824 年，驻印度的英国少校罗伯特·勃鲁士在印度阿萨姆邦沙地耶地区发现了一株高约 13.1 米、直径约 0.9 米的野生茶树。因此，西方学者认为印度是茶的故乡。而事实上，中国才是茶的故乡。中国唐代的《茶经》[①] 中就曾记载，当时中国已发现有直径约 1 米的茶树，高度与英国少校发现的那棵相当，而时间却早了 1 100 年。[②]

茶在中国有四千年左右的历史。在唐代（公元 618~907 年），日本和尚将茶叶带回国，并将茶叶与禅宗结合创造出举世闻名的日本茶道。17 世纪荷兰人将中国饮茶的习惯带到欧洲，并由此传遍欧洲大陆，尤其在英国发展成他们的下午茶文化。19 世纪以前，世界上所有可以饮用的茶叶都来源于中国。[③] 中国人最早发现、栽培和饮用茶叶，并将茶树种子、饮茶方法、栽培技术等传播到

---

① 《茶经》是世界上第一部关于茶的著作，由中国茶道的奠基人陆羽所著。

② 羽叶．中国茶（汉英对照）［M］．合肥：黄山书社，2011：2－4.

③ 叶朗，朱良志著；凯茜译．中国文化读本（德文版）*Blick auf die Chinesische Kultur*［M］．北京：外语教学与研究出版社，2014：225.

世界各地，茶可谓是中国人对世界文化所做的重要贡献之一。

（二）“茶”字的由来

“茶”字在唐朝以前写作“荼”字。“荼”原本的意思是指古时候的一种苦菜。“荼”是形声字，草字头是意符，表明它是草本植物。可是，人们从长期的实践中发现，茶是木本植物而非草本植物，用“荼”指“茶”显然名不副实。

汉代，“茶”字从“荼”中简化出来，有些“荼”字已减去一笔，成为“茶”。中唐时期，读音由“tú”转为“chá”，茶的音、形、义已趋于统一。后来，又因为陆羽所写的茶叶百科全书——《茶经》——广为流传，饮茶的人越来越多，书写“茶”字的频率也越来越高，“茶”的字形进一步得到确立，直至今天。①

（三）中国人与喝茶

中国人的日常生活中有七样必需品：柴米油盐酱醋茶。其中只有茶是饮品，却与食物同样重要。广东人吃早餐总爱去茶楼，泡上一壶茶，点上几件点心，喝完早茶便去上班。对于中国人来说，茶虽然不是饭，不能果腹，但却往往非喝不可。饮茶不仅解渴，还是一种生活方式，超越生理需求上升到了精神层面。②

中国人似乎在任何地方、任何时间都可以饮茶。中国语言学家林语堂曾说，中国人最爱品茶，在家中喝茶，开会时喝茶，劝架讲理也要喝茶；早饭喝茶，午饭后也要喝茶。每当客人来访，主人就会沏上一壶茶来招待客人，这是生活中最常见的一种待客方式。在宴请客人时，主人有时会用酒来表达敬意，而茶是唯一能代替酒的饮品，“以茶代酒”既不失礼仪，又令人愉快。茶不仅是中国人日常和宴客的饮品，也是婚礼、生日、葬礼、祭祀等大型聚会时不可

① 羽叶. 中国茶（汉英对照）［M］. 合肥：黄山书社，2011：9.

② 羽叶. 中国茶（汉英对照）［M］. 合肥：黄山书社，2011：29.

或缺的饮品。[1]

中国很多城市还流行着茶馆文化。中国茶馆与西方咖啡馆有许多相似之处，是休闲娱乐场所。但两者的气氛却有很大的不同，茶馆是热闹的，咖啡馆是安静的。茶馆之所以热闹，是因为其中有很多娱乐活动，如曲艺表演、说唱评书等，人们一边欣赏着表演，一边喝着茶，度过悠闲的时光。现代茶馆的社会功能相当丰富，可饮茶品茗，可小憩休闲，可洽谈商务，还可体验茶文化。[2]

中医理论中茶能治愈多种疾病，因为茶微苦、性凉，含有对身体有疗效的成分。现代药理学对此进行了佐证。饮茶的好处很多，概括起来有以下十四条：

1. 茶能使人精神振奋，增强思维和记忆能力。

2. 茶能消除疲劳，促进新陈代谢，并有维持心脏、血管、胃肠等正常运作机能。

3. 饮茶对预防龋[3]齿有很大好处。据英国的一项调查表明，儿童经常饮茶龋齿发生率可降低60%。

4. 茶叶含有不少对人体有益的微量元素。

5. 茶叶有抑制恶性肿瘤的作用，饮茶能明显地抑制癌细胞的生长。

6. 饮茶能抑制细胞衰老，使人延年益寿。茶叶的抗老化作用是维生素E的18倍以上。

7. 饮茶能延缓和防止血管内膜脂质斑块形成，防止动脉硬化、高血压和脑血栓。

8. 饮茶能兴奋中枢神经，增强运动能力。

9. 饮茶有良好的减肥和美容效果，特别是乌龙茶对此效果尤为明显。

10. 饮茶可以预防老年性白内障。

---

① 羽叶. 中国茶（汉英对照）[M]. 合肥：黄山书社，2011：29－31.

② 羽叶. 中国茶（汉英对照）[M]. 合肥：黄山书社，2011：32－33.

③ 龋（qǔ）

11. 茶叶所含鞣[1]酸能杀灭多种细菌，因此能防治口腔炎、咽喉炎，以及夏季易发生的肠炎、痢疾等。

12. 饮茶能保护人的造血机能。茶叶中含有防辐射物质，边看电视边喝茶，能减少电视辐射的危害，并能保护视力。

13. 饮茶能维持血液的正常酸碱平衡。茶叶含咖啡碱、茶碱、可可碱、黄嘌呤等生物碱物质，是一种优良的碱性饮料。

14. 防暑降温。饮热茶 9 分钟后，皮肤温度下降 1~2 摄氏度，使人感到凉爽和干燥。[2]

对饮茶人士而言，茶的第一大特性是“净化身体”。因为茶树生长在干净的地方，生长环境越干净，茶叶的质量越好。优质的茶树通常长在高山上，被清新空气中的云雾包围。这类茶叶细嫩，采摘时还挂着露水，带着自然新鲜的气息。一杯好茶色泽清澈伴有清香，能起到净化调和身体之功效。

茶的第二大特性是“寻找清闲”。纷繁的世界充满迷惑和争执、厌倦与紧张，对身心造成伤害。一杯清茶能够使人暂时隔离世俗的噪声与混乱。小口啜饮，心灵平静得仿佛身处静夜里的湖泊边，明亮月光投射到世界。一杯茶能顺利地打开通往新世界的道路。

茶的第三大特性是“表达敬意”。中国有习俗，招待客人喝茶以示敬意。客人接茶作为尊敬的一种标志，不管他当下是否口渴，在饮用后都能感到精神焕发、充满活力。在一些聚餐场合，如有人晚到宴席会场，他会以茶代酒自罚数杯。在中国部分地区，会敬客人三次茶，分别表示欢迎他们的到来、热情好客以及美好祝愿。为客人斟茶不仅是好客，还是表达尊敬的举动。[3] 当然，中国也有“端茶送客”的习俗，主人佯装喝茶，其实是委婉地请求来访者离开。

---

① 鞣（róu）

② 李麟. 茶酒文化常识［M］. 太原：北岳文艺出版社，2010：47－48.

③ 叶朗，朱良志著；凯茜译. 中国文化读本（德文版）*Blick auf die Chinesische Kultur*［M］. 北京：外语教学与研究出版社，2014：227－228.

（四） 茶分六色

目前茶叶分类没有统一的方法，鉴于工艺及色泽上的区别，可分为绿茶、红茶、黑茶、青茶、黄茶、白茶六大类。

1. 绿茶

绿茶是不经过发酵的茶，即将鲜叶经过摊晾后直接下到100~200摄氏度的热锅里炒制，以保持其绿色的特点。其制作工艺都经过杀青—揉捻—干燥的过程，依据干燥的方法进行分类，绿茶可分为炒青绿茶、烘青绿茶、蒸青绿茶和晒清绿茶。

绿茶是我国产量最多的一类茶叶，几乎中国的各个产茶区都产绿茶，其花色品种之多居世界首位。绿茶具有香高、味醇、形美、耐冲泡等特点，深受国内外消费者的欢迎。绿茶名品有西湖龙井茶、洞庭山碧螺春茶、庐山云雾茶、黄山毛峰茶、太平猴魁茶、峨眉竹叶青茶、蒙顶甘露茶、信阳毛尖茶、顾渚紫笋茶、普陀山佛茶、都匀毛尖茶、南京雨花茶等。[①]

2. 红茶

红茶与绿茶恰恰相反，是一种完全发酵茶。其品质特点就是红汤红叶，因而叫作红茶。名贵品种有祁红、滇红、宁红等。

红茶与绿茶的区别，在于加工方法不同。红茶加工时不经杀青，而是萎凋，使鲜叶失去一部分水分，再揉捻（揉搓成条，或切成颗粒），然后发酵（使所含的茶多酚氧化，变成红色的化合物）。这种化合物一部分溶于水，一部分不溶于水，而积累在叶片中，从而形成红汤、红叶。红茶主要有小种红茶、工夫红茶和红碎茶三大类别。[②]

3. 黑茶

黑茶属于后发酵茶，是一种中国特有的茶类。所用原料粗糙，加工时堆积

---

① 羽叶. 中国茶（汉英对照）[M]. 合肥：黄山书社，2011：44－45.

② 羽叶. 中国茶（汉英对照）[M]. 合肥：黄山书社，2011：88－89.

发酵时间较长，使叶色呈暗褐色，成茶色泽黝黑，[1] 是我国边疆地区藏族、蒙古族、维吾尔族等少数民族的日常必需品。黑茶原来主要销往边疆地区，像云南的普洱茶就是其中的一种。此外还有湖南的黑毛茶、湖北的老青茶、四川的西路边茶与南路边茶、广西的六堡茶等。[2]

普洱茶是在已经制好的绿茶上浇上水，再经过发酵制成的。普洱茶具有抗衰老、降脂、减肥和降血压的功效，在东南亚和日本很普及。不过真要说减肥，效果最显著的还是乌龙茶。

4. 青茶

青茶也称乌龙茶，是一类介于红茶、绿茶之间的半发酵茶，既有绿茶的鲜浓，又有红茶的甜醇。在六大类茶中，乌龙茶的制作工艺最为复杂、费时，泡法也最讲究，所以喝乌龙茶也被称为喝工夫茶。

半发酵茶，即制作时适当发酵，使叶片稍有红变。因其叶片中间为绿色，叶缘呈红色，故有“绿叶红镶边”的美称。除了具备一般茶叶的保健作用外，还具有抗动脉硬化、减肥健美等功效。乌龙茶中的名品有武夷岩茶、安溪铁观音、凤凰单丛、台湾冻顶乌龙等。

5. 黄茶

黄茶属于轻发酵茶，品质特点为黄叶黄汤、香气清悦、味厚爽口。著名的君山银针茶属于黄茶。黄茶的制法有点像绿茶，不过中间需要焖黄三天。

黄茶在揉捻前或揉捻后，或在初干前或初干后，经过闷堆渥[3]黄，因而形成黄叶与黄汤。根据鲜叶原料的嫩度和大小分为“黄芽茶”（例如，君山银芽、蒙顶黄芽、莫干黄芽）、“黄小茶”（例如，北港毛尖、沩山毛尖、远安鹿苑）和“黄大茶”（例如，霍山黄大茶、广东大叶青）三类。[4]

---

① 羽叶. 中国茶（汉英对照）［M］. 合肥：黄山书社，2011：80－81.

② 秦大东. 黑茶的发展简史［J］. 茶业通报，1983（6）：33.

③ 渥（wò）

④ 羽叶. 中国茶（汉英对照）［M］. 合肥：黄山书社，2011：72－73.

6. 白茶

白茶是一种轻微发酵茶，因其表面布满白色茸毛而得名。主要产于福建的福鼎、政和、松溪和建阳等县。茶叶冲泡后叶片完整而舒展，汤色浅淡或初泡无色，香味醇和。

白茶是我国的特产，其加工方式最简单，有“懒人做茶做白茶”之说。它加工时不炒不揉，只将细嫩、叶背布满茸毛的茶叶晒干或用文火烘干，而使白色茸毛完整地保留下来。白茶的主要品种有银针、白牡丹、贡眉、寿眉等。[①]

（五） 十大名茶

中国名茶在国际上享有很高的声誉。尽管人们对名茶的概念尚不统一，但名茶需有其独特的风格，对茶叶色、香、味、形四个方面进行综合评鉴。中国的“十大名茶”版本众多，众说纷纭。即使不能面面俱到，一些名茶也往往能以其中若干个特色而闻名。本节就以 1959 年全国“十大名茶”评选为标准，分别介绍各种名茶的特征和识别方法。

1. 西湖龙井

西湖龙井是最著名的绿茶品种，同时也是我国第一名茶，因产于杭州西湖山区的龙井而得名。这里依山傍湖，气候温和，常年云雾缭绕，雨量充沛，加上土壤结构疏松、土质肥沃，茶树根深叶茂，常年莹绿。龙井素以“色绿、香郁、味醇、形美”四绝著称于世。茶叶为扁形，叶细嫩，条形整齐，宽度一致，一芽一叶或二叶。色翠略黄似糙米色，香气幽雅浓郁，滋味甘鲜醇和，汤色碧绿黄莹。假冒的龙井多是青草味，夹蒂较多，手感不光滑。

2. 洞庭碧螺春

碧螺春同为著名绿茶品种，产于江苏省苏州市吴县太湖之滨的洞庭山一带。唐朝时就被列为贡品，高级的碧螺春，每公斤干茶需要茶芽 13.6 万 ~ 15 万个。炒成后的干茶条索紧结，白毫显露，色泽银绿，翠碧诱人，卷曲成螺，产

---

① 羽叶. 中国茶（汉英对照）[M]. 合肥：黄山书社，2011：87.

于春季，故名“碧螺春”。常用瓷杯冲泡，杯中白云翻滚，清香扑鼻，品饮时能尝到其特有的花香果味。假冒的碧螺春为一芽二叶，芽叶长度不齐，呈黄色。

3. 黄山毛峰

黄山毛峰是著名绿茶品种，产于安徽黄山，这里山高林密、日照短、云雾多，形成良好的品种。由于新制茶叶白毫披身，芽尖锋芒，且鲜叶采自黄山高峰，因此将该茶取名为黄山毛峰。黄山茶的采制过程相当精细，清明至立夏为采摘期，为了保质保鲜，要求上午采、下午制；下午采、当夜制。制成的毛峰茶外形细扁微曲，状如雀舌，香如白兰，味醇回甘。假茶呈土黄色，味苦，叶底不成朵。

4. 庐山云雾

庐山云雾茶是汉族传统名茶，属于绿茶的一种。最早是一种野生茶，后东林寺名僧慧远将野生茶改造为家生茶。云雾茶始于汉朝，宋代列为“贡茶”。因产自中国江西省九江市的庐山而得名。主要茶区在海拔 800 米以上，由于升温比较迟缓，因此茶树萌发多在谷雨后。由于受庐山凉爽多雾的气候及日光直射时间短等条件影响，形成了叶厚毫多、醇甘耐泡的特点。冲泡后汤色清澈，有兰花、板栗、豆花等自然的清香味。

5. 六安瓜片

六安瓜片产自安徽省六安市大别山一带，成茶呈瓜子形，因而得名“六安瓜片”，是著名绿茶品种，也是名茶中唯一无芽无梗，以单片嫩叶炒制而成的产品，堪称一绝。谷雨前提采的称“提片”，品质最优；其后采制的大宗产品称“瓜片”；进入梅雨季节，茶叶稍微粗老，品质一般，这段时期采制的称为“梅片”。采摘时取二三片叶，求“壮”不求“嫩”。冲泡后水色碧绿，滋味回甜，叶底厚实明亮。假茶则味道较苦，色泽较黄。

6. 君山银针

君山银针属于轻发酵茶，是我国黄茶中的珍品。它产于湖南岳阳洞庭湖的君山，形细如针，故名君山银针。品饮君山银针具有很高的欣赏价值，冲泡时茶具宜用透明的玻璃杯，可以观察到茶芽在热水的浸泡下慢慢舒展开来，芽尖朝上，悬空竖立，在水中忽升忽降，时沉时浮。经过“三浮三沉”之后，最后徐徐下沉杯底，随水波晃动。茶色浅黄，回味甜爽。假冒银针为青草味，冲泡后不能竖立。

7. 信阳毛尖

信阳毛尖是河南省著名特产之一，是我国著名的内销绿茶，主要产地在信阳大别山地区，因茶区的五大茶社产出品质上乘的本山毛尖茶，正式命名为“信阳毛尖”。信阳毛尖具有“细、圆、光、直、多白毫、香高、味浓、汤色绿”的独特风格，饮后回甘生津，冲泡四五次，还能保持长久的熟栗子香。劣质信阳毛尖则汤色深绿或发黄、混浊发暗，不耐冲泡、没有茶香味。

8. 武夷岩茶

武夷岩茶是著名的乌龙茶品种，产于福建北部“秀甲东南”的名山武夷山，茶树生长在岩缝之中。武夷岩茶属于半发酵茶，具有绿茶之清香，红茶之甘醇，是中国乌龙茶之极品。17 世纪，荷兰东印度公司首次采购武夷岩茶，经爪哇转销欧洲各地，成为一些欧洲人日常必需的饮料。冲泡时常用小壶小杯，因香气浓郁，冲泡五六次后余味犹存。茶汤呈橙黄或金黄色，清澈艳丽，叶底软亮，叶缘朱红，叶心淡绿带黄。茶性和缓而不寒，久藏不坏。

9. 安溪铁观音

铁观音同为我国著名乌龙茶之一，主产区位于福建安溪，这里群山环抱，云雾缭绕，年平均气温 15~18 摄氏度。铁观音的名称来源有多种说法，其中一种说法相当符合此茶的特点。相传当年乾隆皇帝细观茶叶，乌润结实，沉重似铁，味香形美，犹如“观音”，便赐名为铁观音。冲泡后具有天然兰花香，汤色清澈金黄，入口微苦，立即转甜，耐冲泡。假茶叶形长而薄，叶泡三遍后便无香味。

10. 祁门红茶

祁门红茶产于安徽祁门县，生叶柔嫩且内含丰富的水溶性物质，八月采摘的茶叶品质最佳，为我国红茶中的珍品。祁门红茶的香气，似苹果香，又似兰花香，清香持久，上品茶更蕴含着玫瑰花香，祁门红茶的这种特有的香味，被国外不少消费者称为“祁门香”。祁门红茶适合加奶加糖调和饮用，最宜于清饮。祁门红茶茶色为棕红色，而假茶一般带有人工色素，味苦涩且淡薄，条叶形状不齐。

（六）茶具

中国人视饮茶为一种蕴含着大智慧的艺术。品茶的滋味似乎很容易，品茶的含义却意味深远。所以，中国人非常重视泡茶过程中水质、温度、茶叶和茶具等各个环节和要素。

如果说水是茶的母亲，那茶具就是茶的父亲。茶具在中国古代泛指制茶、饮茶时使用的各种工具。现代人所说的“茶具”主要指茶杯、茶壶、茶碟、茶勺等饮茶用具。根据材质的不同，可分为陶质、瓷器、紫砂、金属、漆器、竹木、玻璃茶具等。

陶质茶具：是指用陶土烧制而成的茶具，是人类最早制作和使用的茶具之一，最初是粗糙的土陶，然后逐渐演变成比较坚实的硬陶，后来发展为表面敷釉[①]的釉陶，最后发展为紫砂陶。在瓷器茶具出现后逐渐被淘汰，近年又随着茶艺表演的盛行而被重视起来。

瓷器茶具：主要有青瓷、白瓷、黑瓷和彩瓷茶具。浙江龙泉青瓷除具有瓷器茶具的众多优点外，因色泽青翠，用来冲泡绿茶，更有益于表现汤色之美。用它冲泡红茶、白茶、黄茶、黑茶，则易使茶汤失去本来面目，似有不足之处。江西景德镇白瓷茶具远销国外，适合冲泡各类茶叶，加之造型精巧，装饰典雅，其外壁多绘有山川河流、四季花草、飞禽走兽，或缀以名人书法，颇具艺术欣赏价值，使用最为普遍。

紫砂茶具：严格来说与陶质和瓷器茶具不同，介于两者之间。紫砂以陶土为材料，含铁量高，有“泥中泥，岩中岩”之称，产于江苏省宜兴市的紫砂质量最佳。因为在 1 000~1 200 摄氏度之间烧制，所以成品致密，不渗漏。表面

---

① 釉（yòu）

光滑平整，又有细小颗粒的砂质手感，肉眼看不见的气孔使得紫砂茶具拥有很好的吸附性和透气性。耐寒耐热，传热缓慢，不易烫手，也不会爆裂。美中不足的是受色泽限制，用它较难欣赏到茶叶的美姿和汤色。

金属茶具：南北朝时，我国出现了包括饮茶器皿在内的金属器具。隋唐时，其制作技术达到高峰。明代开始，随着茶类的创新、饮茶方法的改变以及陶瓷茶具的兴起，金属茶具逐渐消失。尤其是用锡、铁、铅等金属制作的茶具，用它们来注水泡茶，被认为会使茶味走样，很少有人使用。但金属制成贮茶器具，如锡瓶、锡罐等却屡见不鲜，至今流传于世。此类器具的密闭性要比纸、竹、木、瓷、陶等好，具有较好的防潮、避光性能，这样更利于散茶的保存。

漆器茶具：始于清代，主要产于福建福州一带。福州生产的漆器茶具品种多样，有宝砂闪光、金丝玛瑙、釉变金丝、仿古瓷、雕填、高雕、嵌白银等品种。特别是创造了红如宝石的赤金砂和暗花等新工艺以后，漆器更加绚丽夺目，惹人喜爱。

竹木茶具：隋唐以前，民间多用竹木制作茶具。因为来源广，制作方便，对茶无污染，对人体无害，所以受到人们的欢迎。但缺点是不能长时间使用，无法长久保存，失去收藏价值。到了清代，四川出现了一种竹编茶具，由内胎和外套组成，内胎多为陶瓷类器具，外套用精选慈竹，经多道工序制成粗细如发的柔软竹丝，经烤色、染色，再按内胎形状大小编织嵌合。竹编茶具能保护内胎、减少损坏，同时不易烫手，富含艺术欣赏价值。

玻璃茶具：质地透明，光泽夺目，外形可塑性大，形态各异，用途广泛。用玻璃杯泡茶，茶汤的鲜艳色泽，茶叶的细嫩柔软，茶叶在整个冲泡过程中的上下浮动，叶片的逐渐舒展等，可以一览无遗，可以说是一种动态的艺术欣赏。特别是冲泡各类名茶，茶具晶莹剔透，杯中轻雾缥缈，澄清碧绿，芽叶朵朵，亭亭玉立，观之赏心悦目，别有趣味。玻璃茶杯物美价廉，但容易破碎，比陶瓷烫手。[①]

---

① 李麟. 茶酒文化常识［M］. 太原：北岳文艺出版社，2010：52－70.

如今各种材质的茶具造型丰富，制作精良。茶具经历了一个从无到有，从粗糙到精致的过程，是茶文化的重要载体。

紫砂茶具

二、饮酒

（一）“酒”字的由来

“酒”字的由来有个有趣的传说。远古时期有个人会酿酒，但他酿成的酒总是不好喝。直到有一天，他梦到一位老神仙，老神仙告诉他九天以后的酉时，到马路上取三个人的三滴血放进酒窖，酿出来的酒就好喝了。他按照老神仙的指示去做，分别向一位赶考的书生，一位刚打胜仗回来的将军，以及树下躺着的一个疯子各取了一滴血。回家后他把三滴血放进酒窖，酿出来的酒果然很香。他想到神仙让他“酉”时取“三个人的血”，于是就用三点水和酉这两部分组成了“酒”字。至于神仙让他九天以后去取，这个字就念“九”的音吧。从此以后，饮酒的人也大致分为三种境界，即：书生、将军和疯子。书生举杯文雅，斯斯文文；将军酒过三巡，豪放兴起；疯子则来者不拒，喝完意识全无，不知身处何方。

传说归传说，事实上我国最早的“酒”字出现在甲骨文中，并且有两种写法，一是酿酒器具“酉”的单体象形字；二是中间一个酒瓶，两旁是溢出的酒液，强调容器中饮料的液态性质。后经金文、篆书再到楷书的演变，形成如今以“水”为形符，“酉”为声符的形声字。从写法的不断演变，也可看出酒文化在中国的源远流长。[①]

（二） 中国酒与西洋酒

中国人用谷芽酿造酒和巴比伦人用麦芽做啤酒，差不多同时出现在新石器时代，彼此之间是否有联系却无从考证。中国早在 3 200 年前就有一种用麦芽和谷芽做谷物酿酒的糖化剂，酿成称为“醴[②]”的酒。这种滋味甜淡的酒虽然不叫啤酒，但可以肯定的是它类似现在的啤酒。有文献记载其麦芽的制造过程与现代啤酒工业的麦芽制造过程基本相同。由于后人偏爱用曲酿的酒，嫌醴味薄，以至于这种酿酒法逐渐失传，并被酒曲酿造的黄酒所淘汰。[③]

酒曲酿酒是中国酿酒史的精华，中国人用酒曲酿酒要比欧洲人早 3 000 多年，这得益于中国悠久的农业文明史。酒曲中所生长的微生物主要是霉菌。对霉菌的利用是中国人的一大发明创造。[④] 公元前 138 年张骞出使西域带回葡萄，引进酿酒艺人，中国开始酿造葡萄酒，葡萄酒传入中国比传到法国尚早七八百

---

① 杨姝琼. 解读“酒”字及“酒文化”[J]. 内蒙古电大学刊，2017（06）：33－36.

② 醴（lǐ）

③ 刘勇. 中国酒（汉英对照）[M]. 合肥：黄山书社，2012：56.

④ 刘勇. 中国酒（汉英对照）[M]. 合肥：黄山书社，2012：66.

年。东汉时期发明的蒸馏术，自 18 世纪传入欧洲后，使西方自古以麦芽淀粉糖化谷物，再用酵母菌石糖发酵的传统技术大大提高。中国对世界制酒业的另外一个贡献是煮酒以防止酸败。北宋时期的《北山酒经》中较详细地记录了煮酒加热的技术，而西方各国采用煮酒加热技术的时间比中国晚 700 多年。[①]

（三） 节日酒俗

酒在古代人的日常生活中有着十分重要的作用。每逢节日或喜庆之事，人们都会有举杯饮酒的活动。不同的节日，要饮用不同的酒，于是酒就成为不同节日的载体和体现。例如农历正月初一，以前的普通百姓都要饮屠苏酒或椒酒，因为它们能祛除瘟气，取其平安长寿的寓意，希望家人全年安康；农历五月初五是纪念诗人屈原的端午节，人们会饮用菖蒲酒和雄黄酒，以前人们认为五月是恶月，饮用这些酒可以达到辟邪强身的作用；农历八月十五中秋节，自古就有赏月饮酒的习俗，中秋之酒寓意团圆与思念；农历九月初九是重阳节，为了祈求长寿，古人习惯饮用菊花酒。

这些习俗大多数流传至今，而且与中国特有的药酒文化息息相关。酒本身具有治疗作用，药酒则是将强身健体的中药和食材与酒溶于一体，药借酒力、酒助药势，使得疗效进一步提高。这些传统节日中饮用的酒，如椒柏酒、屠苏酒、菖蒲酒、菊花酒等，都与治疗和预防疾病有关。古代医疗卫生条件比较落后，这些酒俗对预防民间疾病起到积极作用。[②]

（四） 十大名酒

从古至今，白酒在中国人的生活中占有十分重要的位置，是社交、喜庆、送礼等活动中不可缺少的特殊饮品。中国名酒为国家评定的质量最高的酒。十大名酒名字的来历大多与其产地有关，背后都有一段故事，由此也可以看出中国酒文化的博大精深。

---

① 刘勇. 中国酒（汉英对照）［M］. 合肥：黄山书社，2012：1－2.

② 杨姝琼. 解读“酒”字及“酒文化”［J］. 内蒙古电大学刊，2017（06）：33－36.

1. 茅台

茅台酒是因产地而得名。黔北[①]一带水质优良、气候宜人，当地人善于酿酒。人们习惯称这一带为酒乡。而酒乡中又以仁怀县茅台村酿制的酒最为甘洌，谓之“茅台烧”或“茅台春”。由于酒质绝佳，闻名遐迩，世人皆知茅台村出产美酒，他处难以仿制，因此只要提及酒就必说茅台村的酒最好，久而久之就借用茅台地名简称“茅台酒”或“茅酒”。

茅台酒

2. 五粮液

明末清初，宜宾酿酒业已形成一定的规模。“利川永”烤酒作坊老板邓子均选用红高粱、大米、糯米、麦子、玉米五种粮食作为原料，酿造出了香味纯浓的“杂粮酒”。1909 年，宜宾众多社会名流、文人墨客汇聚一堂。席间，“杂粮酒”一开，顿时满屋喷香，令人陶醉。这时晚清举人杨惠泉忽然间问道：

① 黔北：贵州北部地区。

五粮液

“这酒叫什么名字?”在听完酒坊老板的解答后，杨惠泉提议：“如此佳酿，名为杂粮酒，似嫌似俗。此酒既然集五粮之精华而成玉液，何不更名为五粮液?”“好，这个名字取得好。”众人纷纷拍案叫绝，五粮液就此诞生。

3. 泸州老窖特曲

1996 年，国务院将泸州老窖具有 400 年以上窖龄的酿酒窖池群列为国家级重点文物保护单位予以保护。该窖池群始建于明朝万历年间（公元 1573 年），是我国唯一一个建造最早、保存最完好、连续使用至今的酿酒窖池群，被誉为“活文物”，是中华民族的宝贵遗产，具有不可估量的文物价值、社会价值和独特的生产价值。其酿造之酒已成为中国白酒鉴赏标准级酒品，故名“国窖 1573”。

4. 洋河

洋河酒也是因产地而得名。江苏省宿迁市的洋河镇在汉朝时早已是一个酿酒名地。洋河，原名白洋河，白洋河的弯曲处有一清泉，被称为美人泉。白洋河从何而来，美人泉因何形成?民间流传着一个个美丽而动人的故事。洋河大曲正是以美人泉之水酿造而成，酒液澄澈透明，入口鲜爽甘甜，口味细腻悠长，最适合搭配南京盐水鸭、金陵烤鸭等一同品尝。

5. 汾酒

汾酒产于山西省汾阳市杏花村，早在 1 400 多年前此地已有“汾清”这个酒名。当时我国还没有蒸馏酒，史料记载的“汾清”“干酿”等均系黄酒类。而事实上我国的白酒，包括汾酒等名优白酒在内，都是由黄酒演变和发展而来的。晚唐大诗人杜牧的一首《清明》，吟出了千古绝唱：“借问酒家何处有?牧童遥指杏花村。”汾酒也借此声名远播。1915 年，汾酒在巴拿马万国博览会上荣获甲等金质大奖章，为国争光，成为中国酿酒行业的佼佼者。

6. 郎酒

郎酒产地四川古蔺县二郎镇是一块独具神韵的风水宝地。此镇地处赤水河中游，四周崇山峻岭。清朝末年，当地百姓发现在这高山深谷之中有一股“郎泉”，泉水清澈味甜，便开始取郎泉之水酿酒，故名“郎酒”。郎酒厂区附近还“悬挂”着两个天然酒库：天宝洞、地宝洞，洞内冬暖夏凉，常年保持 19 摄氏度的恒温。在洞内贮藏郎酒，可以使新酒醇化老熟更快，且酒的醇度和香气更佳。用天然溶洞贮藏白酒，这在中国白酒生产厂家中是独一无二的。

7. 古井贡酒

古井贡酒以“色清如水晶、香醇似幽兰、入口甘美醇厚、回味经久不息”的独特风格，四次蝉联全国白酒评比金奖。其渊源始于公元 196 年，曹操将家乡亳州产的“九酝春酒”和酿造方法进献汉献帝刘协，献帝对此酒大加赞赏，自此一直作为皇室贡品。南北朝时，在安徽亳州人们发现有一口古井，所酿之酒十里飘香，古井名声大噪，而亳州一带酿酒作坊也如雨后春笋般发展起来。

8. 西凤酒

西凤酒产于陕西省凤翔县柳林镇，始于殷商，盛于唐宋。西凤酒产地是中华民族先祖的定居地区之一，又是民间传说中产凤凰的地方。唐仪凤年间（公元 676~679 年）的一个阳春三月，吏部侍郎裴行俭护送波斯王子回国途中，行至凤翔县城以西的亭子头村附近，发现柳林镇窖藏陈酒香气将五里地外亭子头的蜜蜂、蝴蝶醉倒的奇景，即兴吟诗赞叹。此后，柳林酒以“甘泉佳酿、清洌醇馥”的盛名被列为朝廷贡品。到了近代，柳林酒改名为西凤酒。

9. 董酒

董酒产于贵州省遵义市，魏晋南北朝时期，这里的酿酒业盛极一时，特点是用天然植物酿酒。此酒融汇了 130 多种纯天然中草药，是百草之酒，也是“药食同源”“酒药同源”酿酒起源真正的传承者。“董”字由“艹”和“重”组成，“艹”与“草”同意，“重”为“数量多”之意，而“董”字有正宗、正统、正派、威严、威重之意。“董”字本身的文化内涵与董酒的文化内涵，同时此酒又产于董公寺，这三者之间具有巧妙的联系，1942 年此酒被正式定名为“董酒”。

10. 剑南春

剑南春产于四川省绵竹县，早在唐代就产闻名遐迩的名酒“剑南烧春”。相传李白为喝此美酒曾在这里把皮袄卖掉买酒痛饮，可见此酒非常吸引人。剑南春的前身是“绵竹大曲”。20 世纪 50 年代，蜀中诗人庞石帚在四川大学任教，邀几位忘年之交到家里品尝绵竹大曲时，友人建议将此酒名改为剑南春。三天后，庞先生将写好的“剑南春”三个大字交给了绵竹酒厂，并附上如下解释：“剑南”二字，点出美酒产自剑门雄关之南的绵竹，令人联想起沃野阡陌的天府平原；“春”字乃是古为今用，苏东坡曾经说过唐人酒多以春名，此字催人领略美酒的魅力，给人以春天的启示。①

（五） 酒与书法

从甲骨文、金文里大量出现的“酒”字到如今茅台、五粮液等名酒包装盒上烫制的精美书法字体，中国酒文化与书法艺术自古就有着不解之缘。书法作品《兰亭序》的产生、书写艺术、文辞内容均与酒密切相关。历代众多书法家嗜酒而书，使墨色酒香交融成一道灿烂的风景线。他们的作品因渗透真情实感以及不可复制性，从而具有很高的艺术和历史价值。②

中国古代的书法家、画家大多嗜酒，其一是这些艺术家们往往多愁善感，追求浪漫的生活，具有自己鲜明的个性，因此多借酒兴来达到自己所希望的目的；其二，酒有兴奋中枢神经的作用，可以使人精神亢奋、才思敏捷，能够激发出创作的灵感，“李白斗酒诗百篇”就是流传千古的美丽传说。

东晋“书圣”王羲之（公元 303～361 年）醉酒时挥毫而作的《兰亭序》，是其书法艺术作品的精华，也是中国书法史上最为显赫的书法珍品，有“天下

---

① 李麟. 茶酒文化常识［M］. 太原：北岳文艺出版社，2010：164－175.

② 侯忠明. 中国书法与酒文化［J］. 四川文理学院学报，1999（01）：64－66.

第一行书”的美誉。《兰亭序》的创作动机源于一次野外饮酒游戏——曲水流觞。他和他的诗人朋友们在浙江省兰亭聚会，他们在河水中引出的一条小溪边坐下，溪水上游放着盛满酒的酒器“觞”，让觞顺流而下，停在某人面前，某人就要即兴赋诗一首，赋诗不成就罚喝酒。他们共做 37 首诗，想把当天做成的诗歌整理成集，并推选了德高望重的王羲之来写序言。王羲之感慨万千，乘着酒兴在纸上吐露心中的忧愁与快乐，序言的 324 个字一气呵成，这就是后来影响中国书法千余年历史格局的著名行书文稿《兰亭序》。但等到王羲之酒醒之后，又尝试誊抄 10 遍，却怎么也无法超越他酒醉时创作出的这一版本。[①]

唐代僧人怀素嗜酒，至酒兴起，在寺院的墙壁屏障、衣服器具上写字，人称“醉僧”。他酒醉泼墨，留下了神鬼皆惊的《自叙帖》。李白曾经这样描写怀素：“吾师醉后依胡床，须臾扫尽数千张。飘飞骤雨惊飒飒，落花飞雪何茫茫。”

以“狂草”传于后世的唐代书法家张旭，大醉后灵感骤至，状态极佳，异常兴奋，在庭院内狂呼疾走，一会儿抓笔在手一挥而就，一会儿用头发浸墨而书。醒后看到所写的效果，连自己也觉得神异，于是有了“挥毫落纸如云烟”[②] 的作品《古诗四帖》。[③]

在此列举这些名家例子，并不是为了夸大酒在书法创作中的作用。事实上，他们的激情创作，都是建立在对传统基本功的深刻理解基础之上，单凭嗜酒而写字，并不能写出惊世骇俗的作品的。

（六） 酒与武术

历史上参军的一般都是青壮年男子，军旅的孤寂，战争的残酷，命运无常，生死未卜，造就了他们对酒的心理需求。酒具有减轻紧张情绪、舒缓恐惧心理的作用，能慰藉、麻痹焦躁不安的心灵。而酒能产生刺激情绪、激励勇气

---

① 侯忠明. 中国书法与酒文化［J］. 达县师范高等专科学校学报，1999（01）：64 - 66.

② 唐代诗人杜甫《饮中八仙歌》中赞美张旭书法的诗句，比喻他的书法作品像云烟一样倾泻在纸上。

③ 刘勇. 中国酒（汉英对照）［M］. 合肥：黄山书社，2012：147 - 149.

的催化作用，对于战前鼓舞士气又具有独特的功效。战争年代里积累的攻防技术和自卫本能的升华，构成了中华武术的基础。

人们常说自古诗人皆好酒，其实自古武人也同样好酒。武人好酒是因酒能表现出他们的豪爽气概和尚武精神，借酒寄托他们的情怀。当然，更重要的是酒还成为他们创造超绝武功的“灵浆”。清代著名的傅家拳的创始人傅青主就是在醉中造拳的。

傅山，字青主，别号侨黄，生于1607年，卒于1684年。他是明末清初著名的思想家、诗人、学者、画家和爱国志士，同时他还精于武功。他留下的拳法已经成为一个流派，丰富了我国武术文化宝库。而他的武功更是和他的绘画结合在一起。据传，他作画时总在酒酣之后，独处一室，舞练一番，这才乘兴作画。傅青主在醉中舞拳，进入一种物我两忘境地，然后又将自己的感悟形诸笔墨，因而他的画具有一种山雨欲来的肃杀之气和灵动飞扬的韵味，而他的拳法又具有了一种醉态。酒、画和武术，在他身上融为一体。

“醉拳”是现代表演型武术的重要拳种，其武术招式和步态如醉者形姿，故得其名。醉意醉形曾借鉴于古代的“醉舞”，醉打技法则吸收了各种拳法的攻打捷要，以柔中有刚，声东击西，顿挫多变为特色。作为成熟的套路传承，大约在明清时代。醉拳的流行地区极广，四川、陕西、山东、河北、北京、上海和江淮一带均有流传。

现今流传较广的醉拳可分为两大类：一类是古老醉拳套路，偏重武术实用性；另一类是现代醉拳套路，突出跌扑滚翻、醉形醉态的表现性。不管是哪一类，都要掌握形醉意不醉，步醉心不醉。也就是说，这种“醉”仅仅是一种醉态而非真醉，在攻防中，踉踉跄跄，似乎醉得站都站不稳，然而在跌撞翻滚之中，随着形势发动攻击，使人防不胜防。这就是醉拳的精妙之处。

醉拳看似毫无规矩可言，实则不然。倘若没有武术规矩，其又真正成了醉汉，又何谈武术功夫？醉拳的一举手、一投足都是严守醉拳规矩的，讲究手、脚、步法、眼神的变化、配合与协调，其飘逸、洒脱的动作是经过长期刻苦训练而体现的高度熟练化。

除了醉拳之外，还有“醉剑”。剑术在中国有着悠久的历史，丰富的文化内涵。剑是一种在新石器时代生产的古老兵器，至今仍在大众手中舞练。它曾经是帝王权威的象征，仙家修炼的法器，更成为文人墨客抒情明志的寄托，也是艺术家在舞台上表现人物特点的道具。

就套路而言，剑术有太极剑、太乙剑、八仙剑、八卦剑、七星剑等 30 余种，这些剑术套路有单剑，有双剑；有用长穗，有用短穗；有单手运使的剑，有双手运使的剑；有单人独练的，也有双人对练的，名目繁多，形式不一。

剑术体势中有一种叫作醉剑，这是酒文化浸润的剑术。因为风格独特，深受人们欢迎，尤其适合表演，常常与戏曲、舞蹈艺术相融合。奔放如醉，乍徐还疾，往复奇变是醉剑的运动特点。形如醉酒毫无规律可循，但剑术招式却讲究东倒西歪中暗藏杀招，扑跌滚翻中透出狠手。

剑作为一种武器早已从战场上消失，现在的剑主要是一种健身器械，而剑术已经纯粹是一种和舞蹈相结合的表演项目。而醉剑由于它那如醉如痴、往复多变和动感极强的特点，在舞剑中占据着特殊的地位。①

① 李麟. 茶酒文化常识［M］. 太原：北岳文艺出版社，2010：234－242.

第六章　筵宴文化

筵宴是为了表示欢迎、答谢、祝贺、喜庆等社交活动的需要而举行的一种隆重正式的餐饮活动。筵宴是一种人际交往的礼仪，是人们生活中的美好享受，也是一个国家物质生产发展和精神文明程度的重要标志之一。现代中式筵宴有三个最基本的特征：

1. 以酒为中心安排筵宴。俗话说“无酒不成席”，人们称办宴为“办酒”，赴宴为“吃酒”。酒既可以刺激食欲，又可以助兴添欢。因此，人们注重“酒为席魂”“菜为酒设”的排菜法则。

2. 注重座次安排，讲究进餐礼仪。“设宴待嘉宾，无礼不成席。”注重礼仪是中华筵宴文化非常重要的一部分。正式筵宴的座次安排更是讲究，绝不可以随意就座。一般以长幼、辈分、职位以及社会关系的远近来安排席位。

3. 菜肴品种多样，讲究搭配次序。在菜肴的搭配上，常常冷热、荤素、咸甜、浓淡、酥脆、软硬等相互调和。而且，筵宴菜点的上菜次序也非常讲究。一般来说，现代中式筵宴是三段式的。第一段是“序曲”，传统的、完整的“序曲”内容很丰富、很讲究，包括零食、开胃菜、头汤、凉菜。第二段是“主题”，即筵宴的大菜、热菜，包括头菜、烤（炸）菜、二汤菜、鱼类菜品、荤菜菜品、素菜菜品、甜菜、座汤。第三段是“尾声”，包括主食（面条、米饭等)、时令水果等。

现代中式筵宴的设计遵循“味型搭配合理”和“原材料搭配合理”的原

则，以营造并突出筵宴的主题性、菜肴的独创性、菜名的趣味性。

总之，中国筵宴文化有着悠久的历史、丰富的内涵和较高的艺术观赏价值。中国的筵宴艺术是烹饪艺术精华的体现，是人类文化艺术的重要组成部分。本章将以筵宴礼仪、筵宴餐具、婚宴、百日宴、寿宴、丧宴、家宴、谢师宴和满汉全席中的六宴为主要内容，进一步诠释中国筵宴文化。

一、筵宴礼仪

现代中式筵宴礼仪非常重要，主要包括以下几点。

（一） 座次安排

中式筵宴一般用圆桌，每张餐桌上的具体位次有主次尊卑之分。宴会的主人应坐在主桌上，面对正门就座。同一张桌上位次的尊卑，根据距离主人的远近而定，以近为上，以远为下。同一张桌子上距离主人相同的位次，排列顺序讲究以右为尊，以左为卑。在举行多桌宴会时，各桌之上均应有一位主桌主人的代表，作为各桌的主人，其位置一般应与主桌主人同向就座，有时也可以面向主桌主人就座。每张桌上，安排就餐人数一般在 10 人以内，并且为双数，人数过多，过于拥挤，也会照顾不过来。

（二） 餐桌礼仪

标准的中式筵宴，不论何种风味，其上菜顺序大体相同。通常是首先上冷盘，接着是热炒，随后是主菜，然后上点心喝汤，最后上水果拼盘。当冷盘吃剩三分之一时，开始上第一道菜。宴会上桌数再多，各桌也要同时上菜。

上菜时如果由服务员给每个人上菜，要按照先主宾后主人、先女士后男士或按顺时针方向依次进行。如果由个人取菜，每道菜应放在主宾面前，由主宾开始按顺时针方向依次取菜。越位取菜被视为不礼貌的行为。

正式宴会前，服务员会为每位用餐者递上一条湿毛巾用来擦手。宴会结束时，再递上一块湿毛巾用来擦嘴。正式宴会上，服务员还为每位用餐者准备一条餐巾。它应当铺放在并拢之后的大腿上，而不能把它围在脖子上，或掖在衣领里、腰带上。餐巾可用于轻抹嘴部和手，但不能用于擦餐具或擦汗。

由于中餐的特点和食用习惯，参加中式宴会时，要注意以下几点：第一，上菜后，不要先拿筷，应等主人邀请，主宾拿筷时再拿筷。第二，取菜时要相互礼让，依次进行，不要争抢。第三，取菜时要适量，不要把对自己口味的好菜一人包干；为表示友好、热情，彼此之间可以让菜，劝对方品尝，但不要为他人布菜，不要擅自做主，不论对方是否喜欢，不要主动为其夹菜、添饭，以免让人家为难。第四，不要挑菜，不要在共用的菜盘里挑挑拣拣。第五，取菜时要看准后夹住立即取走，不能夹起来又放下，或取回来又放回去。

（三） 餐具礼仪

中式宴会餐具中最主要的进餐工具是筷子，一般用右手执筷，过高或过低执筷都不太规范。较为规范的执筷姿势是拇指捏按点在上距筷头约占 1/3 处为宜，这样看起来雅观大方，又便于筷子的适当张合使用。民间也有忌讳用餐前用筷子敲空碗，因为在古代，乞丐乞讨时才这样敲碗。另外，把筷子插在盛好的米饭上也是不允许的。因为这是丧礼时敬鬼神的方式，容易让人想到已经去世的人，是不吉利的。

二、筵宴餐具

在餐具方面，现代中式筵宴台面餐具的配置主要包括碗、盘、碟、汤勺、筷子、筷枕、炖盆等。一般来说，盛器的大小要与菜肴的数量相适应，盛具的类型应与菜肴的类型相配合，盛具的色泽应与菜肴的色泽相协调，盛具和盛具应相互配合。

纵观中华文明的数千年历史，中国餐具中瓷器沿用近两千年。作为中华民族对世界文明的伟大贡献，其英文与中国的英文同为“china”。瓷器的特点是质地坚硬、瓷胎洁白、细密轻薄、不吸水等，不仅精致美观、卫生耐用，而且物美价廉、易于贮存。

碗是最常用的餐具之一，它与筷子共同构成了东亚国家进食最基本的形制。此外，盘、盆、碟等也是中华餐具中为数最多的盛食具。现代的盘、盆、碟等与碗一样，在外形上富于变化，圆形、正方形、长方形等在多种场合均可见到。在现代筵宴中恰当搭配使用各种形态的餐具，不仅造型美观，还能烘托气氛、促进食欲。

杯、壶是盛装饮品的专门器具。杯是用来直接饮用液体的，如酒杯用于饮酒，茶杯用于饮茶；壶既可以作为斟茶或斟酒至杯中的间接饮用器，又可以作为独自享用的直接饮用器具。在中式筵宴中，壶大多数为间接饮用器，而杯则是个人饮用器。

经历了漫长的文化进程，筷子、匕、餐刀、餐叉和汤匙等都曾是中华餐具中专门的取食工具。在中华餐饮的演进中，我们也长期使用刀叉、匕等餐具，但最终形成了手与筷子、汤匙结合的取食进餐模式。

筷子是古代中国人民发明的独特的进食用具，象征着古老而悠久的中华文明，是华夏民族饮食文化的结晶。从古至今，筷子可以用金、银、铜、铁、不锈钢等作为材料制成。中式筵宴中的筷子，大多数是用竹木制成。

使用筷子应当正确，并掌握标准的握筷姿势，过高或过低都是不规范的，也不应当变换指法来握筷子。在用筷子夹菜时，不应当在菜肴里来回翻挑与搅拌，更不要用筷子穿刺菜肴。若遇到他人也来夹菜，要注意避让，不要与他人的筷子相碰。不要将筷子含在嘴里或是把筷子当作牙签来使用，这是非常不雅观的。在与人交谈的过程中，不要把筷子当作道具，对人指指点点，这样很不礼貌。用餐结束，要将筷子整齐放回原位，等大家都放下筷子时再一同离席。

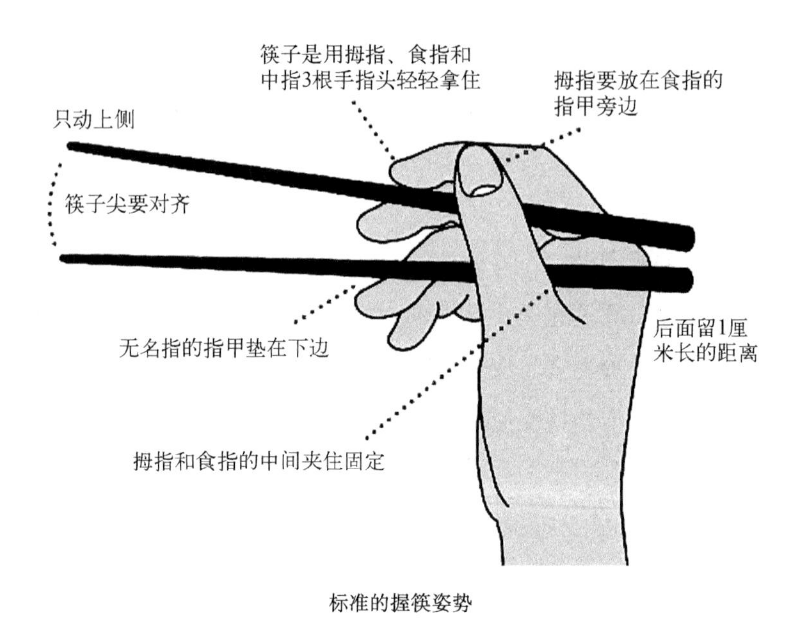
筷子是用拇指、食指和
中指3根手指头轻轻拿住
拇指要放在食指的
指甲旁边
只动上侧
筷子尖要对齐
无名指的指甲垫在下边
后面留1厘
米长的距离
拇指和食指的中间夹住固定

标准的握筷姿势

三、婚宴

婚宴，是新人邀请亲朋好友前来庆祝结婚，以答谢宾客而举办的宴会。参加婚宴，也称“吃喜酒”。现代社会，大多数的婚宴都在酒店举行。在少数地方，仍有在当地家中大院内或空地举办婚宴的。在婚宴开始前，新郎新娘会站在宴会厅门口，迎接宾客的到来，并表示感谢。待宾客到齐、宴会开始前，由司仪主持结婚典礼。司仪会邀请两位新人入场，并向宾客介绍两位新人从相识到相爱的历程。有时邀请双方父母一同来到台前，向新人表达美好的祝愿。在座位安排上，大多数是男方或女方家人亲属、同学同事，年纪相同、互相熟悉的人坐在同一桌。这样，彼此有共同语言，可增加婚宴的喜庆气氛。主桌就座的一般是尊贵的领导嘉宾、双方父母、新人及伴娘伴郎。

在结婚典礼之后，便是酒宴的开始。新人和宾客均开始动筷。稍后，两位新人起身，和伴郎伴娘一起，按照次序，向每桌客人敬酒。在酒宴即将结束时，新郎新娘要站在大厅门口，和将要离去的客人握手道别，有时会说“谢谢光临”“请慢走”之类的话。

结婚之际举办婚宴，在世界大部分国家和地区都能见到。但因各地风俗不同，婚宴的过程和内容也存在着差异。中国幅员辽阔、民族众多，各省、市、自治区的风俗也有很大的差异。

婚宴的菜单有以下几个特点：

1. 菜肴的数目多为双数。我国大部分地区的习俗为“红喜事（即婚宴）”的菜肴数目为双数，“白喜事（即丧宴）”的菜肴数目为单数。如果是八道菜，象征发财。如果是十道菜，象征十全十美。如果是十二道菜，象征月月幸福。

2. 菜名多为吉祥用语以寄托对新人的美好祝福。比如鱼水相依（奶汤鱼圆）、早生贵子（红枣花生桂圆莲子羹）、比翼双飞（珍珠双虾）等。再比如菜肴中有鸡，象征吉祥喜庆；菜肴中有鱼，象征年年有余，并且一般最后一道上。水果一般会选用石榴（多子多福、红红火火）、蜜桃（甜甜蜜蜜）等，但一般不会选用梨（谐音“离”，有“分离”之义）。

3. 预定婚宴菜肴需要考虑宾客的风俗习惯和饮食禁忌，充分尊重宾客的民族风俗、宗教信仰、嗜好和忌讳。比如信仰伊斯兰教的宾客不吃猪肉等。

婚宴菜单参考示例如下：

一彩拼：游龙戏凤（象形冷盘）。

四围碟：天女散花（水果花卉切雕）、月老献果（干果蜜脯造型）、三星高照（荤料什锦）、四喜临门（素料什锦）。

十热菜：鸾凤和鸣（琵琶鸭掌）、麒麟送子（麒麟鳜鱼）、前世姻缘（三丝蛋卷）、珠联璧合（虾丸青豆）、西窗剪烛（火腿瓜盅）、东床快婿（冬笋烧肉）、比翼齐飞（香酥鹌鹑）、枝节连理（串烤羊肉）、美人浣纱（开水白菜）、玉郎耕耘（玉米甜羹）。

一座汤：山盟海誓（大全家福）。

婚宴历来是婚礼过程中非常重要的一部分，由于地域不同、习俗不同，中国婚宴的整个流程呈现多元化的发展趋势。以上内容大多数为中国现代婚宴，传统婚宴流程较为复杂，在这里不再详述。

婚宴

四、百日宴

百日宴，是指婴儿出生之后一百天举行的庆祝仪式。百日宴是中国古代民间的一种风俗，至今仍流行于全国。一般主人会宴请宾客，宾客前来送贺礼，主人也需要回礼，分享爱和喜悦。“百日宴”的主要目的是为了祝福婴儿能够健康成长，长命百岁。在古代，“百日宴”又称“百禄”，“鹿”与“禄”同音，传统祝福为画众多鹿纹相赠，称“百图案多禄图”，象征祝颂之意。

百日宴各有各的规矩。早年民间有“穿百家衣”“吃百家饭”的习俗。这和给孩子起个贱名类似，形式上是沾邻里好友的光。因此有高寿老人的家庭，总是避不了“被借寿”，常要借老人一件衣或一把米，拿去给孩子增寿。如今城市里这种习俗依然常见，虽然只保留了形式，但意思多少还保留着：虽然现代生活状况越来越好，孩子的身体也越来越健康，但大家还是畏于天命，只愿孩子平安。

百日宴的氛围一般以活泼、可爱、温馨、色彩鲜艳为主，如果宝宝是女孩，布置一般以粉色为主，粉红、粉黄、粉紫、白色这些颜色都很好看；如果是男孩，就以实色为主，比如蓝色、白色、黄色。

一般来说，百日宴的菜肴主要分为：凉菜、热菜、汤、饮品和点心。凉菜，例如：钵头醉鸡、香糟小黄鱼、本帮酱鸭、卤水金钱肚、凉拌万年青、水晶鱼冻、翠皮黄瓜、特色素鸡等；热菜，例如：豆豉蒸扇贝王、清蒸鲟鱼、白灼草虾、贵妃蹄髈、水煮鳝背、铁板牛蛙、茶树菇炒牛柳、鱼香茄子、蛋黄炒花蟹、西湖牛肉羹、金牌蒜香鸡、米粉蒸排骨、雪菜炒墨鱼、南瓜百合元宵羹、如意百日面等；汤，例如：神仙老鸭煲、老鹅煲等；点心，例如：上海炒面、寿桃、鲜奶蛋糕等；饮品，例如：橙汁、红酒、酿制酒等。

从百日宴菜肴的数量来看，一般为吉利和谐的双数，一桌十个人，至少八道菜，但一般是十二道菜最为合适，表示月月圆满。十六道菜也是很常见的，如果菜的分量少，可以点十六道菜。但尽量秉承不浪费的原则。

从百日宴菜肴的质量来看，好吃、量足、丰盛、新鲜等都是评价菜肴是否合格的标准。另外宴会一般要求有荤有素，有鱼有肉，有凉菜也有热菜，最好有海鲜，这样会更有档次感，菜品也更丰盛，只要这些都满足了，嘉宾会对此赞不绝口的。

从百日宴菜肴的创意来看，由于是为宝宝举办的宴会，可以更为童趣一些，那么菜单也需下点功夫来体现孩子的童真可爱，比如各种可爱造型的包子、米奇形状的米饭或者小熊造型的蛋糕，都很可爱，让人觉得很有童心。

从百日宴菜肴的禁忌来看，一般情况下，不能点乳猪做的菜肴，因为民间有一种说法是乳猪就是猪宝宝，在宝宝的宴会上食用乳猪寓意不好，给人一种不吉利的感觉，所以不能点。还有含有鱼卵的鱼也不要吃，会比较残忍，毕竟妈妈十月怀胎才生了小宝宝。有的地区是属鸡的宝宝在百日宴上不能出现鸡肉等禁忌。中国民族众多、地域广袤，各地习俗也差异较大，在举办百日宴之前，最好能够结合当地习俗来点菜。

百日宴

相比婚宴，庆祝百日宴的人并没有那么多。不是每个家庭的婴儿到了100天都要举办百日宴，也有相当一部分家庭并不举办百日宴。其中有一部分原因是婴儿还太小，免疫力较差，理应尽量减少接触成年人，减少到人群密集的地方去。因此，如果举办百日宴或参加百日宴的时候，一定要注意以下几点：

1. 宝宝如果一同赴宴，众人最好不要搂抱宝宝，亲吻宝宝更应避免。如果那样，宝宝会缺乏安全感，并且容易感染病菌或疾病。为了宝宝的身心健康，对于众人要求抱宝宝的举动可以善意拒绝。

2. 最好不要擅自给宝宝喂食，宝宝在1岁之前不能摄取含糖、含盐的食物。而且宝宝牙齿没有长出，宴会上的大多数食物都无法咀嚼，特别是食用坚果类食物，可能会出现噎住、卡住的现象，对于只有百天的宝宝来说会有生命危险。父母最好给宝宝自带食物和玩具，转移宝宝对满桌食物的注意力。

3. 在宴会厅内，一般会开着空调，夏天温度较低，冬天又太过暖和，所以宝宝进出宴会厅一定要适度增减衣物，以防受凉感冒。

4. 宴会结束后要及时休息，宝宝的精力毕竟是有限的，无休止的宴会宝宝肯定挺不住，最好不要打乱宝宝的作息时间。

五、寿宴

寿宴，是指为老人祝寿的宴席。其风俗因地而异，但都会大宴宾客。

很久以前，人们原本不过生日。儒家的“孝亲”理论认为“哀哀父母，生我劬[①]劳”。越是遇到生日，越应该想到父母生养自己的艰辛，生日这天要静思反省，缅怀双亲的辛劳。因此“古无生日称贺者”。然而在南北朝时期，已有过生日的仪式。《颜氏家训》中有每年过生日要设酒食庆贺的记载。有趣的是，庆贺生日与不庆贺同样是出于“孝亲”的观念，不庆贺是为了体悟父母的辛苦，而庆贺则是为了娱亲。唐代，民间普遍以做生日为乐事，设酒席、奏曲乐，对生日当事人祝吉祝寿。自此，纯粹以祝寿祝吉为目的、以酒宴乐舞为形式的生日庆贺习俗一直流传至今。自宋代起，过生日“献物称寿”的送礼之风日渐兴盛，生日馈赠礼仪沿袭至今，已成为过生日的一项重要习俗。

古代称老年人为“寿”，“寿”意味着生命的长久。出于孝道，每逢老人生日，子女必举办隆重的祝寿仪式，大摆寿筵，广邀亲朋，登堂拜寿，以示孝心。对寿诞的重视，充分体现了中华民族尊老敬老的传统美德。

现在一般以 50 岁或 60 岁为分界线，过此年龄的才能做寿。多数地方自 60 岁开始，俗语有“祝六十大寿”之说。也有的地方在 40 岁就开始做寿。

寿诞前一日为“暖寿”。这天，一是寿星的女儿和女婿带着礼品回到娘家，同寿星和家人一起共进晚餐；二是在家里布置好寿堂，陈列子女及亲朋的寿

---

① 劬（qú）

礼，做好第二天祝寿仪典的最后准备工作。第二天的寿诞日在家设宴款待亲朋好友，一般要沿袭古代遗风行叩拜礼。这种形式大多流行于农村，尤其是偏远地区农村。城市大多数在酒楼、宾馆设宴庆寿。在庆寿地摆设和布置寿堂，寿庆仪式与酒宴同时进行。仪式程序可多可少，一般是古今结合、中西结合。

寿宴是寿礼的重要一环，主家往往设宴席，款待来客。寿宴的菜单中有鸡鸭鱼肉、海鲜，但必不可少的是“长寿面”。贺寿的来客都会携带寿礼，诸如寿桃、寿糕、寿面、寿烛、寿联、寿彩、万年伞等。这些礼品中但凡能缀饰、点画图案的，一般都要加上一些象征长寿的图案。

一般情况下，先招待客人食用鸡蛋、茶点、长寿面。寿星本人一般不在正堂入座，而是找几个年龄相仿的老者作陪，在里屋另开一席。菜肴品种多多益善，取多福多寿之兆。寿宴菜品多扣“九”“八”，如“九九寿席”“八仙菜”。除了寿面、寿桃、寿糕等面点外，还有白果、松子、红枣汤等。菜名较为讲究，如“八仙过海”“三星聚会”“福如东海”“白云青松”等。鱼菜少上，不上西瓜盅、冬瓜盅、爆腰花等。长江下游一些地区，每逢父（或母）亲 66 岁生日，出嫁的女儿要为之祝寿，并将猪腿肉切成 66 小块，形如豆瓣，俗称“豆瓣肉”，红烧后，盖在一碗大米饭上，连同一双筷子一并置于篮内，盖上红布，送给父（或母）亲品尝，以示祝寿。肉块多，寓意老人长寿。

寿宴的会场布置大多数金碧辉煌，餐桌上铺着红色的台布，中间摆放葱绿鲜艳的美丽植物，大厅周围张灯结彩，红红的灯笼点缀其间，显示寿堂的喜庆热闹气氛。墙壁上贴着由儿孙亲自写在红纸上的寿字图案，或者大型松柏、仙鹤祝寿图，突出了环境的主题，热闹且美观。有的宴会会场贴上一副“天增岁月人增寿，春满乾坤福满门”内容的寿联，寿联簇拥着一个大大的红色“寿”字。“寿”字下有两张寿星椅，它的左右两端各摆放一大一小两张礼案（即方桌），礼案上摆放福、禄、寿三星，鲜寿桃等祝寿物品。寿堂下面摆放四个装有无数小气球的大气球和一个扎有红色彩带的大礼包，用来放置儿孙给老寿星贺寿的礼物。背景音乐多为步步高、祝寿歌等。寿宴过后，寿星本人或由儿孙代表，向年高辈尊的亲族贺客登门致谢，俗称“回拜”。

无论是婴孩的周岁生日，青少年、成年人的平时生日，还是老年人的寿诞，庆贺仪式的繁简根据家庭经济状况差别较大，庆贺仪式的程序也因地域不同而各具特色。

寿诞礼仪在少数民族中也很盛行，并且在某些方面有其独特之处。比如壮族举行寿礼时，晚辈要用猪肉、鸡来祭祖先。行礼之后，大家还要簇拥着老人唱“祝寿歌”。

另外，寿诞有一些吉祥物，例如寿桃、鹤、龟、松、五瑞图、“寿”字吉祥物、寿面、月季、葫芦、寿石、南山、猫、蝶和鹿等。

在寿宴上，无论是由西方传入的生日蛋糕，还是中国传统的寿面、寿桃、寿糕等，都包含着祝福寿星健康长寿与幸福吉祥的美好愿望。

下面是某酒店的寿宴菜单，仅供参考：

福如东海宴

福星高照（锦绣大拼盘）

吉庆满堂（生灼游水虾）

福寿双全（脆奶海皇卷）

金鸡贺寿（脆皮烧鸡）

老当益壮（金沙焗肉蚧）

福如东海（清蒸大海斑）

春色满园（北菇扒芥胆）

寿比南山（蟠桃寿包）

长寿富贵（鹅肝酱烧伊面）

鸿运年年（万寿果炖双雪）

环球鲜果（生果拼盘）

寿宴

六、丧宴

丧宴，是指家里人为缅怀已经去世的人而举办的宴席，通常会邀请亡者的亲朋好友前来参加。“白事”是办理丧葬的一种称呼，中国民间白事礼仪分为送终、报丧、入殓、守铺、搁棺、居丧、吊唁、接三、出殡、落葬、居丧，它是带着美好祝愿的。因此中华文化中所讲的“红白喜事”，就是指结婚和去世两类喜事。丧宴是“白事”的重要环节之一。

在参加丧宴时，应该注意切忌大声喧哗，手机等发声设备应该关机或调成振动模式。在宴席上注意与亲朋好友聊天的话题不要太过令人兴奋。如果亲属前来敬酒，也应该礼貌回应，并道“节哀顺变”。

如果亡者德高望重，家族兴旺，是全福；年龄在 60 岁以上，即为满寿、全寿；一个人无病痛地去世，自然而去，是善终。亡者若是汇集全福、全寿、善终而去，主人可以大摆宴席招待亲朋好友和邻里乡亲加以庆祝。人们以这种看似喜庆的形式，将悲痛减少，仍然要继续健康快乐地生活下去。因此，这样的酒宴往往办得隆重热闹。

另外，主人往往会回赠客人毛巾、碗等。毛巾作为洗涤织物，寓意涤荡晦气，还可以用来给前来帮忙的亲朋好友擦汗擦泪。而且，毛巾属于大量消耗的日用品，价格不高，对于要承担葬礼及白宴的主家来说，也算是减少开支。如果是参加自然而终的或百岁老人的丧宴时，主人通常会把碗送给客人，据说这样可以带来“寿气”，可以保佑宾客长命百岁。

丧宴菜单菜肴数目一定为单数，忌双数。举办丧宴的时候用九样东西，象征对主人家去世一人深表遗憾。古时丧宴菜单中一般避免大鱼大肉，以素食居多以悼念死者。丧宴中必须有羊肉，因为羊懂哺乳之恩，有跪着吃奶的举动，所以食羊肉表孝心。在古代，还会用豆制品和素菜所做的豆腐饭来祭奠死者并招待以金钱、物质、人力来帮助自家办理丧事的亲朋好友。古人认为豆腐是长寿食物，并且豆腐的谐音为“都福”，以祈愿家人和来客能福寿双全。如今已经演变成了各种美味佳肴和美酒，不再限于豆腐和素菜。

由于中国民族众多，各地风俗略有不同，因此真正想了解一个地方的丧宴文化，最好的方法就是进行实地考察。例如广东白事酒席，先上三牲（猪肉、鱼、鸡），一般不必拘泥于形式。客家酒席的最后一道菜是青菜。关于酒席用酒，红事喜庆用酒不受限，但白事酒一般只用黄酒和白酒。饮白事酒时，忌猜拳行令、喧哗及饮醉闹事。江浙一带参加丧宴可以准备白包给主人以示哀悼，金额不能双数、整数，通常可以再加一元放在白包里。

七、家宴

家宴是在家里举行的宴会，通常是限于家庭范围、规模较小、菜肴相对比较丰富的饮食聚会。相对于正式宴会而言，家宴最重要的是要制造亲切、友好、自然的气氛，使赴宴的宾主双方感觉轻松、自然、随意，彼此增进交流，加深了解，促进信任，往往不做特殊要求。为了使来宾感受到主人的重视和友好，基本上要由女主人亲自下厨烹饪，男主人充当服务员；或男主人下厨，女主人充当服务员，来共同招待客人，使客人产生宾至如归的感觉。

家宴的安排原则如下：

1. 考虑季节的特点。配置家宴时，必须根据季节的变化来调整菜肴的内容，使菜肴品种能与季节相适应。外界气温的改变，在一定程度上可以影响人体的热量消耗和对食物的消化吸收，以及人们的饮食心理状态，因此在烹制家宴时，要考虑季节的变化。

2. 注意冷热搭配和口味的变化。在烹制家宴时，既要有凉菜也要有热菜，不论是冷菜或热菜，应尽量不要出现两种或两种以上同一种口味的菜肴。一般2~4 人凉菜 2 道，5~7 人凉菜 4 道，8~10 人凉菜 6 道。

3. 注意菜肴的品种、色泽和营养的搭配。调配家宴时要荤素搭配，刀工要有变化，菜肴的色泽要五彩缤纷。要考虑一料多菜，如“一鱼三吃”“一鸡两吃”等，使菜肴既经济实惠，又丰富多彩。

4. 了解和掌握客人的饮食特点。如逢年过节时，人们进食各种食物较多，一般要求油腻少、质量高、平时不易吃到的新品种。所以在配置菜肴时，应少大鱼大肉，菜肴的数量不宜过多。另外，也要注意客人的生活习惯和口味特点。例如，一般江浙、上海等地区喜甜食、偏清淡，而四川、贵州等地区则喜好带酸辣味的菜肴。

5. 要注意现烹现食。除了冷菜、全鸡、全鸭等短时不易烹熟的菜肴要事先准备外，一般的热炒菜以现烹现食为佳。因为食物和调味品都含有各种营养成分，经加热后会有不同程度的变化，趁热食用既可保持菜肴的营养成分，其口味也最佳。如菜肴凉后再回锅加热，则色、香、味、形都会产生变化，风味尽失。

隆重的家庭宴会是比较讲究上菜顺序的，一般情况为先凉后热，先菜后点，先咸后甜，先炒后烧，先优质（特色风味菜）后一般，先菜肴后面点、水果。具体顺序如下：第一是摆冷盘以佐酒，可以是什锦拼盘，或四双拼盘、四三拼盘等，也可以采用一个花色拼盘，周围再围上四个、六个或八个小单盘。第二是热炒，一般要求采用滑炒、生炒、炸、溜、爆、烩等多种烹调方式，以达到菜肴的口味和外形多样化的要求。第三是大菜，用于家宴的大菜大多数是由整只、整块、整条的原料烹制而成，装在大盘或大汤碗中上桌的菜肴。它一般采用炖、焖、蒸、烤等烹调方式。第四是甜菜，甜菜在一席菜中所占比例较小，一般只有一个或两个，一般用拔丝、蜜汁、炸、蒸等烹调方法烹制而成，多数是趁热上席，但在夏令季节也可以不用热食品种，而改用冷食品种。第五是点心，用于家宴的点心有糕、团、面、酥、包、饺等品种，具体的种类与成品的粗细，取决于家宴规格的高低。第六是汤，可以根据客人的喜好，做番茄蛋汤、紫菜鲜虾汤等。第七是水果。

现代家庭如举行西餐宴席，则上菜顺序为冷菜、汤羹、热菜，餐前应上开胃酒和鸡尾酒。西餐应最先上冷菜，上菜方式有两种，一种是客人入席之前，先将冷盘摆放在桌上；另一种是待客人入席坐定之后再上冷菜。冷菜包括拼盘、沙拉、黄油、果酱及面包等，冷菜上完后应上汤，接着再上热菜。如果热菜有几道，则要注意肉、鱼、禽、虾与各种蔬菜的合理搭配。待客人吃完菜肴后，再上各类甜点和水果，最后上红茶、咖啡等饮料。同时应送上牛奶和方糖，供部分客人按需取用。

当菜端上桌时，主人应该大致介绍一下这道菜的风味或特色，如果客人对某一道菜特别感兴趣时，主人还可以简单介绍此菜的烹饪方法。通常，在餐桌上客人有主次长幼之分，主人应先请主宾或长者品尝。当客人出现相互谦让、不肯下筷的情况时，主人可站起来用公筷为客人分菜。分配顺序一般是先分给长者，然后按照就座的次序依次分。

上菜的时机必须掌握适当，便于客人饮酒和品尝菜肴。冷菜客人吃三分之二左右，就可以上热菜了。这时客人一边饮酒一边品尝菜的味道。客人饮酒适量时，大菜上完，随即上一道甜菜、甜汤，让客人清清口，换换口味，接着再上饭菜、面点，随后上汤。待客人们酒足饭饱，剩余菜点全部撤掉，清理桌面，随上水果，宴席结束。

如果要参加宴会，那么就需要注意，首先必须把自己打扮得整齐大方，这是对别人也是对自己的尊重，还要按主人邀请的时间准时赴宴。除酒会外，一般宴会都请客人提前半小时到达。如果由于某些原因在宴会开始前几分钟到达，不算失礼，但迟到就显得对主人不够尊敬，非常失礼。

当走进主人家或宴会厅时，应首先跟主人打招呼。同时，对其他客人，不管认不认识，都要微笑点头示意或握手问好；对长者要主动起立，让座问安；对女宾举止庄重，彬彬有礼。

入席时，自己的座位应听从主人或招待人员的安排，因为有的宴会主人早就安排好了。如果座位没定，应注意正对门口的座位是上座，背对门的座位是下座。应让身份高者、年长者以及女士先入座，自己再找适当的座位坐下。

入座后坐姿端正，脚踏在本人座位下，不要任意伸直或两腿不停摇晃，手肘不得靠桌沿，或将手放在邻座椅背上。入座后，不要旁若无人，也不要眼睛直盯盘中菜肴，显出迫不及待的样子，可以和左右同席客人简单交谈。

用餐时应该身着正装，开席前不要脱外衣，更不要中途脱衣。一般是主人示意开始后再进行，就餐的动作要文雅，夹菜动作要轻。要把菜先放到自己的小盘里，然后再用筷子夹起放进嘴里。要小口进食，不要倚靠在桌上，以免碰到邻座。不要在吃饭、喝饮料、喝汤时发出声响。如要用摆在同桌其他客人面前的调味品，先向别人打个招呼再拿；如果太远，要客气地请人代劳。如果用餐时需要剔牙，要用手或手帕遮掩。

喝酒的时候，一味地给别人劝酒、灌酒，特别是给不胜酒力的人劝酒、灌酒，造成别人醉酒，都是失礼的表现。

如果宴会没有结束，但你已用好餐，不要随意离席，要等主人和主宾餐毕先起身离席后再离开。

现代家宴菜肴不仅讲究美味，同时也很注重健康。大多数家宴均使用公筷和公匙，少数实行分餐制。这样不仅干净卫生，减少浪费，也便于适应个人口味，让客人感到备受尊重。

家宴待客，最重要的是让亲朋好友感受到最真实的浓情与关怀。喜庆或节日聚餐，亲朋好友的兴趣在“聚”而不是“餐”。互相之间有很多话要谈，如果主人只忙于做菜，忙于“服务”，即使菜肴做得再好，也会使客人有遭到冷落之感，不易吊起大家胃口。此外，提倡“主随客便”，做到既热情劝食，又不勉为其难，喝酒随量，吃菜随意，不强加于客人。

八、谢师宴

谢师宴，也称升学宴，一般是指学生在高中、本科或研究生毕业之际邀请老师一起聚餐，以表达学生对恩师的感激之情。

谢师宴一般是学生的自愿行为，选择场地的时候，应考虑大多数学生的家庭经济情况。因为办谢师宴主要的目的是感谢恩师对学生多年来的培养，而不是为了拼排场、护面子。谢师宴的菜单很重要，菜品最好兼顾每位老师及同学的口味。有些可以指定菜单的谢师宴，还会考虑到将菜谱与恩师联系起来，以增加喜庆氛围，博得师生的好感。菜品是谢师宴成功的一大因素，最好提前了解一下老师的个人喜好。在谢师宴开始之前，学生应尽量邀请每一位老师，包括不同课程的老师、朝夕相处的辅导员等。

在谢师宴现场，学生应注意文明礼貌，切忌吸烟、说脏话等不文明行为。现场气氛应当较为活跃，学生们金榜题名、学业有成或是找到了如意工作，都是值得高兴的事情，可以适当喝点红酒或白酒活跃气氛，但要注意节制，不可暴饮。一般情况下会有学生向老师敬酒的环节，可以趁着敬酒的机会跟老师说说心里话。谢师宴是学生生涯一段经历的节点，学生可以借此机会与老师拍合照，留下这值得纪念的一瞬间。

谢师宴如果仅仅是吃顿饭，意义并不大。学生应该精心准备各个环节，以多种形式来感谢恩师，例如让学生说说自己的心里话，和老师聊聊天，学生代表和老师代表发言等。

常见的谢师宴有以下几种：同窗惜别宴（宴同学）、望子成龙宴（宴亲友）、金榜题名宴（宴师长）。

谢师宴菜单一份供参考：

锦绣前程——经典卤水拼盘

寒窗苦读——豆豉肉片炒苦瓜

一丝不苟——青椒鳝丝

知恩图报——鹌鹑蛋烧甲鱼

天光云影——老南瓜绿豆排骨汤

斗志昂扬——五香脆皮鸡

勇往直前——糯米蒸排骨

欢聚一堂——水果拼盘

青出于蓝——上汤浸时蔬

年年有余——清蒸鳜鱼

尊师重道是中国几千年来的传统美德，谢师宴一是可以表达对老师的感激之情，二是可以培养学生的感恩之心。但过于奢侈的谢师宴会给师生情笼罩上物质化和庸俗化的阴影，并不值得提倡。

九、满汉全席

满汉全席据说兴起于清代，集满族与汉族菜点的精华，包括满族的烧烤、涮锅、火锅，还包括汉族的扒、炸、炒、熘、烧等烹饪技术。满汉全席上菜一般至少 108 种，其中南方菜 54 道：30 道江浙菜、12 道闽菜、12 道广东菜；北方菜 54 道：12 道满族菜、12 道北京菜、30 道山东菜。乾隆甲申年间（约 1764 年）李斗所著《扬州书舫录》中记载有一份满汉全席食单，是关于满汉

满汉全席

全席的最早记载。

满汉全席共有六宴，分三天吃完，均以清宫著名大宴命名。入席前，人们要上二对香，桌面上摆着茶几和手碟，还有四鲜果、四干果、四看果和四蜜饯等。入席后，先上冷盘，随后上热炒菜、大菜、甜菜。合用全套粉彩万寿餐具，配以银器，富贵华丽，用餐环境典雅隆重。席间有专门的乐师弹奏音乐伴宴，使人们美食与美音俱享。

满汉全席既有宫廷菜肴之特色，又有地方风味之精华，菜点精美、礼仪讲究、取材广泛、用料精细。烹饪技术精湛，富有地方特色。满汉全席以东北、山东、北京、江浙菜为主。所谓“满汉全席”中的珍品，大部分是黑龙江地区特产（或出产）。后来福建、广东等地的菜肴也依次出现在巨型宴席上。根据百度百科的资料介绍，满汉全席中的六宴分别为：

1. 蒙古亲藩宴

此宴是清朝皇帝为招待与皇室联姻的蒙古亲族所设的御宴。一般设宴于正大光明殿，由满族一、二品大臣作陪。历代皇帝均重视此宴，每年循例举行。受宴的蒙古亲族更视此宴为大福，对皇帝在宴中所例赏的食物十分珍惜。

2. 廷臣宴

廷臣宴于每年正月十六日举行，由皇帝亲点大学士、九卿中有功勋者参加，因此能够参加宴会的人都感到十分光荣。宴会场所设于奉三无私殿，宴时遵循宗室宴的礼节。全部都用高椅，赋诗饮酒，每年循例举行，蒙古王公等也都参加。皇帝借此施恩来联络下臣，而同时又是廷臣们功禄的一种象征。

3. 万寿宴

万寿宴是清朝帝王的寿诞宴，也是内廷的大宴之一。后妃王公、文武百官，都以进寿献寿礼为荣。宴席上名食美味不可胜数。如遇大寿，则庆典更为隆重盛大。衣物首饰，装潢陈设，乐舞宴饮一应俱全。光绪二十年十月初十是慈禧六十大寿，于光绪十八年颁布上谕，寿日前几个月，筵宴已经开始。仅事前江西烧造的绘有万寿无疆字样和吉祥喜庆图案的各种釉彩碗、碟、盘等瓷器就达 29 170 余件。整个庆典耗费白银近 1 000 万两，规模在中国历史上是空前的。

4. 九白宴

九白宴始于康熙年间。康熙初定蒙古外萨克等四部落时，这些部落为表示投诚忠心，每年以九白为贡，即：白骆驼一匹、白马八匹，以此为信。蒙古部落献贡后，皇帝设御宴招待使臣，称为九白宴。每年循例而行。

5. 节令宴

节令宴是指清宫内廷按固定年节时令而设的筵宴。例如：元日宴、元会宴、春耕宴、端午宴、乞巧宴、中秋宴、重阳宴、冬至宴、除夕宴等，均按节次定规，循例而行。满族虽有他们固有的食俗，但是入主中原后，在满汉文化的交融中和统治的需要下，大量接受了汉族的食俗。由于宫廷的特殊地位，食俗规定逐渐详尽，其食风又与民俗和地区有着紧密联系，所以腊八粥、元宵、粽子、冰碗、雄黄酒、重阳糕、乞巧饼、月饼等食品在清宫中一应俱全。

6. 千叟宴

千叟宴始于康熙时期，盛于乾隆时期，是清宫中规模最大、与宴者最多的盛大御宴。康熙五十二年在阳春园第一次举行千人大宴，康熙帝席间赋诗《千叟宴》一首，因此“千叟宴”为该宴名。乾隆五十年于乾清宫举行千叟宴，与宴者 3 000 人。嘉庆元年正月再次在宁寿宫皇极殿举办千叟宴，参与宴会的人有 3 056 人，即席赋诗 3 000 多首。后人称千叟宴是“恩隆礼洽，为万古未有之举。”

满汉全席的菜式以当时为准，现在略有改动，就不再列举具体菜式。

如今，我们几乎没有机会品尝满汉全席，而且有些菜品已不合宜，但可以从菜单中挑出几样，通过网络或书籍查阅其做法，来品尝不同地方的精华菜品，品味中华饮食文化的瑰宝。

21 世纪，人们生活条件和消费观念逐渐发生了变化，人们在饮食上追求创新、营养、卫生，促进了筵宴向更高境界发展，进入了筵宴的繁荣创新时期。传统筵宴正在不断改良，创新筵宴也在大量涌现。同时，还积极引进了西方宴会的形式，形成了中西合璧的画面。中国筵宴充分体现出了东方文化中祥和、佳美、新颖等风格特点，这是历代祖先留给中华民族的瑰宝。我们应该在学习中国传统文化的同时，批判继承、创新发展！

参考文献

[1] 爱心家肴美食文化工作室. 新编家常菜谱 2 [M]. 青岛：青岛出版社，2009 年.

[2] 陈光新. 中国筵席宴会大典 [M]. 青岛：青岛出版社，2001 年.

[3] 楚丹. 书法 [M]. 合肥：黄山书社，2011 年.

[4]《超值典藏》编委会. 中华民俗大全集 [M]. 北京：中国画报出版社，2012 年.

[5] 都大明. 中华饮食文化 [M]. 上海：复旦大学出版社，2011 年.

[6] 杜莉，姚辉. 中国饮食文化 [M]. 北京：旅游教育出版社，2016 年.

[7] 呼志强. 最有趣的民俗知识 [M]. 北京：中国纺织出版社，2012 年.

[8] 柯玲. 中国民俗与文化 [M]. 北京：北京大学出版社，2017 年.

[9] 李麟. 茶酒文化常识 [M]. 太原：北岳文艺出版社，2010 年.

[10] 刘怡，芮鸿. 活在丛林山水间：云南民族采集渔猎 [M]. 昆明：云南教育出版社，2000 年.

[11] 刘勇. 中国酒（汉英对照）[M]. 合肥：黄山书社，2012 年.

[12] 马健鹰. 中国饮食文化史 [M]. 上海：复旦大学出版社，2008 年.

[13] 孔建民. 生活小窍门大全 [M]. 南京：江苏科学技术出版社，2003 年.

[14] 乔娇娇. 中国饮食 [M]. 合肥：黄山书社，2014 年.

[15] 瞿明安，秦莹. 中国饮食娱乐史 [M]. 上海：上海古籍出版社，2011 年.

[16] 王学泰. 中国饮食文化简史 [M]. 北京：中华书局，2010 年.

[17] 吴澎等. 中国饮食文化 [M]. 北京：化学工业出版社，2013 年.

[18] 叶昌建. 中国饮食文化 [M]. 北京：北京理工大学出版社，2011 年.

[19] 谢定源. 中国饮食文化 [M]. 杭州：浙江大学出版社，2008 年.

[20] 萧帆. 中国烹饪百科全书 [M]. 北京：中国大百科全书出版社，1992 年.

[21] 姚伟钧，刘朴兵，鞠明库. 中国饮食典籍史 [M]. 上海：上海古籍出版社，2011 年.

[22] 叶朗，朱良志著；凯茜译. 中国文化读本（德文版）：*Blick auf die Chinesische Kultur* [M]. 北京：外语教学与研究出版社，2014 年.

[23] 俞为洁. 中国食料史 [M]. 上海：上海古籍出版社，2011 年.

[24] 羽叶. 中国茶（汉英对照）[M]. 合肥：黄山书社，2011 年.

[25] 张恩来. 家庭厨房万事通 [M]. 南京：江苏科学技术出版社，2004 年.
[26] 张景明. 中国北方游牧民族饮食文化研究 [M]. 北京：文物出版社，2008 年.
[27] 张景明，王雁卿. 中国饮食器具发展史 [M]. 上海：上海古籍出版社，2011 年.
[28] 赵凡禹，水中鱼. 30 岁前要学会的 33 堂礼仪课 [M]. 上海：立信会计出版社，2010 年.
[29] 赵建民，金红霞. 中国饮食文化概论 [M]. 北京：中国轻工业出版社，2011 年.
[30] 赵荣光. 中华饮食文化 [M]. 北京：中华书局，2012 年.
[31] 赵荣光. 中国饮食文化概论 [M]. 北京：高等教育出版社，2003 年.
[32] 赵荣光. 中华酒文化 [M]. 北京：中华书局，2012 年.
[33] 诸葛文. 三天读懂五千年中华民俗 [M]. 北京：中国法制出版社，2014 年.
[34] 陈涓. 地理环境对我国饮食文化的影响 [J]. 福建教育学院学报，2003 (4).
[35] 胡小平. 重阳酒 [J]. 国际金融，1999 (11).
[36] 侯忠明. 中国书法与酒文化 [J]. 四川文理学院学报，1999 (1).
[37] 李军. 中韩传统节日饮食文化比较研究——以春节和中秋为例 [J]. 亚太教育，2016 (29).
[38] 刘昌浩，李秀敏. 中国传统小吃的现存问题以及改进措施 [J]. 商情，2018 (7).
[39] 吕晓敏，丁骁，代养勇. 中国八大菜系的形成历程和背景 [J]. 中国食物与营养，2009 (10).
[40] 蒋志英. 中国八大菜系及第九菜系 [J]. 文史精华，2013 (5).
[41] 满来. 北京小吃的品种及分类 [J]. 时代经贸，2012 (5).
[42] 庞乾林，林海，阮刘青，李西明. 中国稻米文化和现代成就 [J]. 中国稻米，2004 (3).
[43] 秦大东. 黑茶的发展简史 [J]. 茶业通报，1983 (6).
[44] 万建中. 中国菜系的形成 [J]. 人民周刊，2016 (6).
[45] 王文辉，佟伟，贾晓辉. 我国冻梨生产历史、产业现状与问题分析 [J]. 保鲜与加工，2015 (11).
[46] 杨中俭. 上海本帮菜 [J]. 四川烹饪，2014 (11).
[47] 杨姝琼. 解读“酒”字及“酒文化” [J]. 内蒙古电大学刊，2017 (6).
[48] 杨琳. 中秋节的起源 [J]. 寻根，1997 (8).
[49] 杨健，孙大庆. 东北粘豆包发酵面团中细菌多样性研究 [J]. 黑龙江八一农垦大学学报，2015 (10).
[50] 杨择. 中秋节起源 [J]. 江淮，2004 (9).
[51] 周旺. 论西南桂、滇、黔三省区少数民族小吃文化的区域个性 [J]. 饮食文化研究，2005 (3).
[52] 张妍，张橙. 东北地区少数民族饮食文化研究 [J]. 人文高地，2014 (1).
[53] 刘慧. 上海老饭店本帮菜文化传承研究 [D]. 上海：华东师范大学，2015.

Basic Course of
Chinese Culinary Culture

Contents

Chapter 1 Everyday Cuisine

Food is one of the basic necessities of life. As an old saying goes, "Man is iron and food is steel, they are inseparable; skipping one meal will leave a man feeling starving." More profoundly, food is the essential precondition for existence and development of human. As Karl Marx put it, "Mankind has to sustain life in order to 'create history'. Thus, food, clothing, shelter and other necessities become the requisites for that end. Therefore, the first activity in human history was creation of these necessities, i.e., creation of the material life."① Chinese people have been fully aware of the importance of food since ancient times. As Confucius once said, "Desire for food and sex is human nature." "The Conveyance of Rites" in *The Book of Rites* also says, "Food and sex are the major physical desires of human beings." i.e., eating is one of the basic attributes of mankind. Over the centuries, the Chinese people have developed a unique culinary culture. Dr. Sun Yat-sen once made the following comment: "China has fallen behind the other countries in terms of advancement of modern civilization. Only in development of its cuisines has China consistently outperformed the countries of the civilized world."②

---

① *Selected Works of Mark and Engels* (volume 1)[M]. Beijing: People's Publishing House, 1972: 79.

② Sun Yat-sen (author). Mu Zhi et al. (commentary). *The International Development of China*[M]. Shenyang: Liaoning People's Publishing House. 1994: 5.

1.1 Dietary Structure

A description of dietary structure of Chinese people can be done both vertically and horizontally. Vertically, or diachronically, dietary structure of Chinese people has displayed different characteristics during each historical period due to migration of ethnic groups, higher productivity and spread of species. Horizontally, or synchronically, regions with distinctive culinary culture have been formed on the vast territory of China as a result of differences in geographical conditions, climatic conditions and ethnic composition.

1. Evolution of dietary structure

Diachronically, some scholars divide Chinese culinary history into the following periods: the prehistoric period; the pre-Qin period; the Qin and Han Dynasties period; the Wei, Jin, Southern and Northern Dynasties period; the Sui, Tang and Five Dynasties period; the Song-Liao-Jin-Yuan period; the Ming and Qing Dynasties period; the modern and contemporary period. Each period features unique dietary structure due to differences in productivity and crops grown.

1) Prehistoric period

The basic feature of diet in prehistoric times is "RU MAO YIN XUE" which means eat animal flesh raw and drink its blood literally. The so-called "MAO" which means hair literally in fact is not the real hair of fur in the present sense, but refers to vegetable plants. Collection of wild plants and fishing & hunting were the two primary sources of food during the prehistoric period.

As fishing and hunting could not serve as a stable source of food during the prehistoric period, collection of wild plants became the most important source. Long time exposure to the environment of the barbarous wilderness enabled people to identify which plants were edible while which plants were poisonous and not edible. Legends about Shennong reflect people's food seeking experience during this period. As described in *Huainanzi·Xiuwuxun*, Shennong "tasted a varieties of herbs as well as water from many springs in order to teach the common people which herbs were edible and water of which springs was drinkable. During the process, he even found seventy herbs which were poisonous in the course of a day." The story shows that the people of the prehistoric period made tremendous efforts to find food and their hard work paid off. In the traditional ballads of ethnic minority groups living in today's Yunnan, there are also lines describing how the ancient people collected wild plants as food: "climbing high mountains and picking berries, peeling the berries at home and let's call it a day. Going to the tree-filled valley and finding bunches of wild vegetables, cooking them at home for today's meal. Carrying a bundle of firewood back home and building a strong fireplace fire, lying down for a night's sleep."① So far, a great amount of collected food remains have been found in archaeological excavations. For example, large quantities of water chestnut remains were found at 8,000-year-old Bashidang site in Li County, Hunan Province. According to Mr. Pei Anping, "If converted, the quantity of water chestnut remains is comparable to the quantity of rice remains having been found so far."② Wild plants having been found at prehistoric sites in China include: water chestnut; hemp seed; wild rice; eucalyptus; chestnut tree; plum; apricot plum; apricot; plum tree; cherry; peach; persimmon; wild jujube; elm seed; walnut; hazelnut; pine nut; pear; hawthorn; muskmelon; soybean; olive etc.

Despite the difficulties, fishing and hunting remained one of the primary sources of food during the prehistoric period. With the improvement in productivity, people were able to obtain more food from fishing and hunting activities. People started fishing and hunting in the remote past. At Shaanxi Lantian site, stone balls

---

① Liu Yi and Rui Hong. *Living In Jungles and Mountains: Hunting and Gathering Life of Ethnic Groups in Yunnan* [M]. Kunming: Yunnan Education Publishing House. 2000: 15.

② Pei Anping. Rice Remains of Pengtoushan Culture and Rediscussion on Prehistoric Rice Agriculture in China [J]. *Agriculture Archaeology* (Issue 1, 1998): 197.

for hunting were found. Tools for hunting were also found at other sites such as Shaanxi Liangshan site and Henan Sanmenxia site. Bows and arrows were invented during the late Paleolithic period. Stone arrowheads were found at 30,000-year-old site in Shuo County, Shanxi, indicating a leap forward in hunting techniques during that period.

As discovered from archaeological excavations conducted within the territory of China, what the prehistoric people had caught from fishing and hunting activities can be divided into the following categories: (1) Animals such as deer, boar and wild ox. At the Beijing Cave Man site, a great deal of skeletal remains of deer were found. Large quantities of skeletal remains of spotted deer, ox, boar, tiger and elephant were also found at Longtan site in He County, Anhui, a site contemporaneous with Cave Man site. At Majiabang cultural site in Jiaxing, Zhejiang, it was found that "there were animal bones weighing about 1,000 kilograms in the 50 square meters T1 and T2. At the bottom of the lower layer, animal bones were scattered everywhere reaching a thickness of 20 centimeters to 30 centimeters."[①] (2) Aquatic animals such as fish and clam. Regions rich in water resources were home to aquatic animals whose remains were found at Hemudu site in Zhejiang. "During excavation, remains of a wide variety of aquatic animals including fish, terrapin and mussel were found and it's hard to determine the exact quantity of remains of each individual type. We cleaned up a small portion of turtle remains which included over 2,000 pieces. Remains of soft-shelled turtle were also impressive in terms of quantity. Due to corrosion, only bright white mussel remains were found on the ground of the site. Skeletal remains of aquatic animals such as fish, terrapin and mussel were found in many burned pottery kettles. Fragments of fish bone remains were clearly seen in dog's poop."[②] (3) Insects such as pupa, ant

---

① Zhejiang Cultural Relics Administration Committee. Excavation of Neolithic Sites of Majiabang Culture in Zhejiang Jiaxing[J]. *Archaeology* (Issue 7, 1961): 351.

② Zhejiang Cultural Relics Archaeology Research Institute. *Hemudu: Report on Archaeological Excavation of Neolithic Sites*[M]. Beijing: Cultural Relics Publishing House. 2003: 198.

and locust. As these foods were not easy to preserve, remains of very few of them have been found from excavation. However, according to anthropological research, these protein-rich foods must have constituted a major part of the diet of the prehistoric people.

With cultivation of plants and domestication of animals, farming and animal husbandry have gradually replaced plant collection, fishing and hunting as the primary sources of food. Crops grown during the early periods of China included millet, broomcorn millet, rice, hemp, wheat etc.

Millet, broomcorn millet and wheat are major crops grown in the areas north of the Yangtze River. Millet and broomcorn are indigenous to China. The climate of the Yellow River Basin is suitable for the cultivation of millet and broomcorn millet as the two crops require a short period of growth and are tolerant to drought. Widely planted in the Yellow River Basin, the two crops were the staple food during this period. Wheat was introduced from West Asia between 5,000 BC and 3,000 BC During this period, boiling was the major food preparation method. Without decrustation, wheat kernels were hard to be cooked thoroughly so that people may feel bloated after eating. As a result, wheat based foods had not been widely consumed before the invention of mill. Rice is a crop originating in China. Currently, most sites of unearthed rice are in the middle and lower reaches of the Yangtze River with the 6,000-7,000 years old Hemudu site offering the largest concentration. Hemp was the source of textiles during the prehistoric period and hemp seeds were edible. Remains of hemp seed were found at archaeological sites in Henan and Gansu showing that hemp seeds were consumed as food during the prehistoric period.

Pigs, dogs, sheep, cattle and chickens were among the first to be domesticated, thereby becoming an important part of the diet of Chinese people. Pig remains were commonly seen livestock remains at prehistoric sites in China. Pig bones were unearthed at Nanzhuangtou site in Xushui County, Hebei, which dated back approximately 10,000 years. Report on excavation of Nanzhuangtou site indicated that the pig bones unearthed might belong to domestic pigs. Remains of dogs, sheep, cattle and chickens were also found, although in small amounts.

2) Pre-Qin period

With the improvement of farming tools and methods as well as the increase in labor force, agricultural planting had replaced food collection as the major food production form during the Xia and Shang Dynasties. Animal husbandry had greatly developed and raising livestock in pens was widely practiced. Both species being raised and livestock products had been increased.

Crops grown during the pre-Qin period were almost the same as those grown during the prehistoric period including millet, broomcorn millet, wheat, barley, rice, beans (soybean) and hemp. However, thanks to the invention and wide use of metal tools, both planting areas and crops yields were significantly increased. Pair tillage and intensive cultivation also helped improve land utilization ratio and increase crop yields. More wheat and beans were grown during this period. Regionally speaking, millet, broomcorn millet, wheat and beans were mainly grown in the North while rice was grown in the South. The invention and refinement of mill drove the cultivation of wheat.

The development of animal husbandry during the Shang Dynasty was most significant. Some believed that Shang was a nomadic people so that it had experienced seven migrations from Shang Tang to Pan Geng. In the records of oracle bone inscriptions, there are such words as hay, herding and fold which depict the practice of raising livestock in pens, indicating the development of animal husbandry during this period. *Guanzi · Qingzhongwu* says, "Shang government called on their people to build pens and raise cattle and horses for the benefits of the people", which describes the development of animal husbandry during the Shang Dynasty. *Guanzi · Lizheng* also says, "If six kinds of animals are raised by every household and vegetables and fruits are grown, our country will enjoy prosperity." The six kinds of animals refer to horses, cattle, sheep, chickens, dogs and pigs. Horses and cattle were mainly used during labor while sheep, chickens, dogs and pigs were mainly kept for food.

As for daily diet, the scholar-officials ate more meat. The "meat eaters" mentioned in *Zuozhuan* refers to the members of the upper class. The average people followed vegetarian diet. Only during the prosperous years could the elderly eat meat. As described in *Mengzi · King Hui of Liang*, "If chickens, dogs and pigs are raised properly and in due time, those aged 70 and above would have meat for food."

3) Qin and Han Dynasties

The biggest change during the Qin and Han Dynasties was the wider cultivation of wheat in the North, which was mainly driven by the improvement of water conservancy technology and the invention and refinement of mill. Wheat is divided into barley and wheat. The seed covering of wheat is too hard to be consumed as grains. However, ground wheat is sticky suitable for making wheaten food. The seed covering of barley is soft and ground barley is not sticky capable of being boiled and eaten as grains. Therefore, barley was most widely cultivated during this period. Mill was invented in the late Spring and Autumn period and the early Warring States period. Some scholars believe that only water mill was available during that period. Water mill could only produce fluid-like food instead of dry powder. After improvement of mill, wheat became more suitable to be consumed as food. Therefore, wheat which was better quality and more freeze resistant replaced barley which was less tolerant to drought and produced less yields, and wheat was widely cultivated in the Central Plains region. As in the old days, rice was mainly cultivated in the South. In the meantime, rice began to be grown in some areas in the North where conditions were favorable for cultivation.

Soybeans were probably the staple food of the ordinary people. As recorded in *The Book of Han · Huozhizhuan*, "The rich decorate their houses with silk. For them, there is still meat and millet food left even after their dogs and horses are fed. On the other hand, the poor wear course clothes, eat beans and drink unboiled water." That is, the rich eat meat and millet food while the poor only have beans as their food. Soybeans are easy to grow and tend to produce high yields.

There were two sources of meat during the Qin and Han Dynasties: animal husbandry in grassland area; livestock farming in farming area. During the Qin and Han Dynasties, the Central Plains region may capture a huge number of livestock from the wars with the Huns including cattle, horses and sheep. In farming areas, livestock such as pigs, sheep and chickens were raised. Sometimes, cattle and horses were also raised although beef and horse meat were not consumed in large quantity. Poultry such as ducks and geese were also raised. Staring from the late Spring and Autumn period, aquatic products were harvested from artificial breeding and rearing, which further developed during the Qin and Han Dynasties. As the Qin and Han Dynasties expanded their territories, foreign foods such as grape, buckwheat, clove and peas were introduced into China offering more food choices for Chinese people. In addition to traditional cooking methods such as steaming and boiling, new methods such as roasting were also employed. With the improvement of fermentation technology, preparing unleavened dough was not the only way of making wheaten food. Wheaten food can also be made after fermentation.

4) Wei, Jin, Southern and Northern Dynasties period

The Wei, Jin, Southern and Northern Dynasties period was an important period for the formation of the Chinese nation featuring unprecedented people-to-people exchanges between different ethnic groups. Along with the wider cultivation of wheat and rice, new foods and cooking methods were introduced.

Familiar items such as cucumber, eggplant and pepper were introduced into China during this period. According to Mr. Li Jiawen, a culinary culture specialist, "Cucumber has two ecotypes in China. Southern ecotype cucumber was introduced directly from Southeast Asia and is now mostly seen in south China. This type of cucumber required warm and humid climate. The feature of being short-day plant may facilitate staminate flower differentiation and yield short and thick fruits without noticeable edge thorns. Northern ecotype cucumber was introduced from the Central Asia about 2,000 years ago. After being nurtured in north China, its ecological characteristics had undergone significant changes. It can grow in the continental climate in the North, a region known for its dry climate and often drastically changing temperatures. Except the early-maturity varieties, this type of cucumber is long-day plant which differentiated pistillate flowers under long-day sunlight. Its fruits were long and thin with noticeable edge thorns."

Originally grown in India and Southeast Asia, eggplant was introduced into China along with Buddhism and gradually became a frequently seen dish on the dining-table. Eggplant was first widely grown in the South.

5) Sui, Tang and Five Dynasties period

Sui, Tang and Five Dynasties period was one of the periods where foreign cultural exchange was the most dynamic. During the powerful Tang Dynasty, foreign cultural exchange went beyond the Western Regions creating a "Han culture circle" which exerted its influence on surrounding areas. Dietary structure during this period was also varied.

The most significant change during this period was related to crops. Rice became another important food source along with millet and wheat. This was because the South had been developed since the Wei, Jin, Southern and Northern Dynasties period and the political and economic center moved to the South. Since the Wei, Jin, Southern and Northern Dynasties period, the ethnic minorities flocked to the North forcing the residents of the Central Plains to move to the south of the Yangtze River. As a result, the labor force in the South increased leading to development of the South. When An-shi Rebellion took place, the North was plagued by fire of war while the South remained relatively safe. Large populations moved to the south of the Huaihe River further driving the development of the South. Thus, the South gradually became the economic and political center of China.

With regard to vegetable cultivation, spinach was introduced into China during this period and soon became widely grown in the country. Originally grown in today's Nepal, spinach became widely cultivated because it was freeze resistant and easy to grow. People of the Tang Dynasty had gained adequate knowledge of spinach. As Meng Shen and Zhang Ding recorded in their *Dietetic Materia Medica*, "(spinach) is cold in property, slightly toxic, beneficial to five internal organs, freeing intestines and warming stomach, relieving alcoholism; best for those eating medicinal powder; serving as a dietary balance for the northerners who normally eat meat and wheaten food; cold for the southerners who normally eat fish, turtle and rice; not consumed in large quantity due to cold effects on intestines; long period of consumption leading to physical weakness and inability to walk and backache; not to be eaten along with fish which may lead to cholera and vomiting."

Mustard was also introduced during this period. Other items introduced into China during this period included: lettuce; almond; pistachio; gack fruit; fig; watermelon; horsebean; date; sea palm; date palm; olive; spices such as clove and cumin.

6) Song-Liao-Jin-Yuan period

During the Song-Liao-Jin-Yuan period, rice became the No.1 crop in China. As the saying goes, "If Suzhou and Huzhou have a bumper harvest; the country will be safe from the threat of hunger." This was the result of population migration to the South since the Wei, Jin, Southern and Northern Dynasties period. On one hand, people having moved to the South brought advanced production techniques and tools which helped improve productivity in the South. On the other hand, increased population led to more wasteland being reclaimed. Along with the favorable natural conditions of the South, the South eventually became the "granary on the earth". The southern migration of the people of the North also helped wheat become the No. 2 crop in the South.

Sorghum was a newly emerging crop during this period. During the Jin-Yuan period, sorghum became an important food source in the North. During the Song

Dynasty, cabbage became a major vegetable and was largely grown in the South. After years of cultivation, nappa cabbage was produced during the Song Dynasties growing in both the North and the South. The best nappa cabbage was from Yangzhou.

During this period, the nomads moved to the Central Plains region transforming farmland into pastures. The change had its influence on dietary structure where mutton and milk products became popular. Not only did the nomads of the Liao, Jin and Yuan Dynasties have a preference for mutton, the Han people also loved eating mutton. During the Northern Song Dynasty, mutton was the favorite food of both the imperial family and the common people.

7) Ming and Qing Dynasties period

The Ming and Qing Dynasties period maintained the pattern where rice was grown in the South and wheat was grown in the North. During this period, American crops such as maize, sweet potato and potato were introduced into China.

Maize was introduced into China during the reign of Jiajing Emperor of the Ming Dynasty and became widely grown in the country during the reign of Qianlong Emperor of the Qing Dynasty. Due to a huge surge in population during the Qing Dynasty, the Qing government encouraged the cultivation of maize in order to meet the growing demand for food.

Sweet potato was introduced into China during the reign of Wanli Emperor of the Ming Dynasty. By the end of the Ming Dynasty, Fujian and Guangdong had become famous sweet potato production regions. After the Ming Dynasty, sweet potato had become popular in China and was widely grown across the country.

Potato was introduced into China at the end of the 18th century. Back then, most areas consumed potato as vegetable just like today. When a plague struck or in some impoverished areas, potato was eaten as a staple food.

Other items coming to China during the Ming and Qing Dynasties period were:

chili; peanuts; onion; tomato; summer squash; pumpkin; pineapple; sunflower; mango; balsam pear etc.

2. Regional characteristics of dietary structure

Exchange between ethnic groups and migration of ethnic groups as well as geographical conditions had a great impact on the evolution of culinary culture. Synchronically, Mr. Zhao Rongguang divided Chinese culinary culture into the following areas: "After a long period of development and integration, a number of culinary culture areas have formed in China. They include northeast culinary culture area, Beijing and Tianjin culinary culture area, culinary culture area in the middle reaches of the Yellow River, culinary culture area in the lower reaches of the Yellow River, culinary culture area in the middle reaches of the Yangtze River, culinary culture area in the lower reaches of the Yangtze River, north central culinary culture area, northwest culinary culture area, southwest culinary culture area, southeast culinary culture area, Tibetan Plateau culinary culture area, vegetarian culinary culture area."① Mr. Zhang Jingming revised the list by deleting the vegetarian culinary culture area. According to Mr. Zhang Jingming, "Instead of being a regional culinary culture area, vegetarian culinary culture area is a special area having been found in all the other culinary culture areas."② He also changed "north central culinary culture area" into "northern grassland culinary culture area". According to Mr. Zhang Jingming, "north central culinary culture area is dominated by a culinary culture created by the nomads in the North. Ethnic groups living this area rely on nomadism and animal husbandry as a way of life and mode of production."③

---

① Zhao Rongguang and Xie Dingyuan. *Introduction to Culinary Culture*[M]. Beijing: China Light Industry Press, 2006: 49.

② Zhang Jingming. *A Study on Culinary Culture of Nomadic People in Northern China*[D]. Beijing: Minzu University of China. 2003: 3.

③ Zhang Jingming. *A Study on Culinary Culture of Nomadic People in Northern China*[D]. Beijing: Minzu University of China. 2003: 3.

1) Northeast culinary culture area

According to Mr. Zhao Rongguang, the northeast culinary culture area includes Heilongjiang, Jilin, Liaoning and the portion of Inner Mongolia Autonomous Region which is close to the northeast. This culinary culture area features cold and long winters, fertile soil, rich water resources and large area of grassland and mountain forest. Crops in this area require a short period for growth. Ethnic groups inhabiting this area include Han, Mongolian, Manchu and some other ethnic minority groups. The Han people come from Shandong, Henan, Hebei and other places.

Wheaten food is the staple food in the northeast region including jiaozi, steamed bun and steamed stuffed bun. As the northeast region has abundant meat and fish resources, animal protein comprises a large part of its diet than the Central Plains region and other places. Vegetables eaten in the northeast include napa cabbage, radish, potato and other types which are easy for long time storage. People in the northeast consume more soybean products than people of other regions. Major soybean products include soft beancurd, dried beancurd, soybean sprout and mung bean sprout. Frozen food also comprises a large part of the diet of the people of the northeast, a region known for its coldness. A rich variety of frozen food is eaten such as frozen jiaozi, frozen beancurd and frozen fruits. In the northeast, spring and summer are short and winter is long and cold so that fresh vegetables sometime are not available. As a result, pickled vegetables and sun dried vegetables are prepared in order to ensure adequate supply of vegetables during the long and cold winter. Pickled vegetable from the northeast is famous in the country.

2) Beijing and Tianjin culinary culture area

Since the Song and Yuan Dynasties, Beijing and Tianjin had been the political center of the country. Jurchen people, the Mongolians, the Han people and Manchu

Frozen fruits

had established their respective capital here and as a result, foods from across the country can be found in this region. Foods eaten by the northern nomads such as beef and mutton and foods from the South became a part of the local diet. Despite the food diversity, wheaten food is still the staple food of Beijing and Tianjin as the region is located in the North. Wheaten foods include steamed bun, steamed stuffed bun and different flavors of noodles. Meat consumed most includes pork, beef and mutton. In spring and summer, various kinds of vegetables in season are available. In the cold winter, vegetables easy for storage are eaten such as napa cabbage, potato and radish.

3) Northern grassland culinary culture area

The northern grassland culinary culture area, with Inner Mongolia at its center, covers Xinjiang, Gansu, Ningxia, Shaanxi, Shanxi, Hebei, Liaoning, Jilin and part of Heilongjiang. Ethnic minority groups living in this area include the

Mongolians and the Evenki who rely on nomadism and animal husbandry as a way of life and mode of production.

Meet and cheese were the staple foods in this area. In his *Jean de Plan Carpin Histoire Des Mongols*, Jean de Plan Carpin, the 13th century missionary, described, "They (the Mongolians) held large numbers of livestock including camels, cattle, sheep and goats. They possessed many stallions and mares. I don't think other places in the world have the same number of horses. They don't have pigs or livestock for farming." "They ate everything that was edible such as dog, wolf, fox and horse...They didn't make bread or had plants, vegetables or other foods. They only had meat."①

As transport has become more convenient in recent years, rice, wheaten food and vegetables comprise a larger part of the local diet in addition to meat. Crops hold a more prominent position than meat in some farming areas.

4) Northwest culinary culture area

With Xinjiang at its center, the northwest culinary culture area covers Gansu, Qinghai and part of Tibet. The regions' population comprises mainly the Uyghurs, Kazaks, Hui and Mongolians. Equal attention has been paid to the development of agriculture and animal husbandry. Wheat and rice are the major crops and mutton is widely consumed. Fruits such as grapes and cantaloup are famous around the world.

Nang

---

① Geng sheng, He Gaoji. *Jean de Plan Carpin Histoire Des Mongols /The Journey of William of Rubruk to The Eastern Parts*[M]. Beijing: Zhonghua Book Company. 1985: 30, 41.

Due to religious belief, beef, mutton and dairy products are the staple foods. Wheaten food such as Nang and noodle is also seen in everyday diet. Rice is sometimes eaten. Unlike the past, vegetables comprise a larger part of the local diet.

5) Culinary culture area in the middle reaches of the Yellow River

The culinary culture area in the middle reaches of the Yellow River covers Shaanxi, Shanxi, Gansu, Henan, Qinghai and part of Ningxia. This region is the major production area of wheat and wheaten food is the staple food of the local diet. The local people are good at making noodles such as the oat noodles and hand-pulled noodles of Shaanxi, sliced noodles of Shanxi, hand-pulled noodles of Gansu and stewed noodles of Henan, all of which are famous delicacies in China. Other staple foods are small hard flour pancake of Shaanxi and baked pancake of Henan.

Pork is consumed most in this area while beef and mutton are eaten most in some areas inhabited the Hui ethnic group. The diet of the Han people has been greatly influenced by that of the Hui ethnic group. For example, beef and mutton are quite popular in Zhengzhou and Xi'an. Seasonal vegetables instead of stored vegetables and pickled vegetables are part of the everyday diet.

6) Culinary culture area in the lower reaches of the Yellow River

With Shandong at its center, the culinary culture area in the lower reaches of the Yellow River also covers neighboring provinces including Henan, Anhui and Jiangsu. This area produces all kinds of food grains. Rice was grown as the major crop in earlier days. The cultivation of maize receives more attention after the mid-Qing Dynasty and finally maize replaced rice as the major crop grown in this area. Thin pancake made from ground food grains is the staple food of the local people which is eaten along with various kinds of paste and vegetables. The coastal area in the east is rich in seafood resources so that seafood comprises a part of the local diet. Wheaten food such as steamed bun, steamed stuffed bun, steamed twisted roll and noodle which are made from wheat flour is also part of the local diet.

7) Culinary culture area in the middle reaches of the Yangtze River

The culinary culture area in the middle reaches of the Yangtze River covers parts of Hunan, Hubei and Jiangxi. Thanks to its favorable natural conditions and rich water resources, this area has been an important grain producing area in history, as demonstrated by the saying, "if Hunan and Hubei have a bumper harvest, the country will be safe from the threat of hunger." Rice is the staple food while wheat is also grown in some places on a small scale. Maize and sweet potato have been grown on a larger scale sine the Qing Dynasty. Home to the Xiangjiang River, the Dongting Lake and the Dongjiang Lake, this area produces aquatic products in large quantities. Steamed Fish Head with Diced Hot Red Peppers and Stone Pot Fish of Hunan are two famous dishes. Pork is widely consumed in this area. The cured meat of Hunan is a famous delicacy across the country.

8) Culinary culture area in the lower reaches of the Yangtze River

The culinary culture area in the lower reaches of the Yangtze River covers Jiangsu, Zhejiang, Anhui and Shanghai. This region is the birthplace of rice. Large quantities of rice remains were found at 6,000–7,000 years old Hemudu cultural site. Today, paddy rice is still the major crop and boiled rice is the staple food in this area. Home to a vast network of rivers, this area in the lower reaches of the Yangtze River produces aquatic products in large quantities. Fish, shrimp and crab are available all year round. The coastal area is rich in seafood resources and seafood comprises a part of the local diet. People in this area consume more bamboo shoots than the people in the North. Pork and duck are widely consumed. Vegetables and fruits are available throughout the year.

9) Southeast culinary culture area

The southeast culinary culture area covers Fujian, Guangdong, Hainan, Guangxi and Taiwan. Located in subtropical zone, this area does not have large seasonal temperature difference and produces large quantities of fruits and vegetables throughout the year. The coastal area is rich in seafood resources. Rice is the staple food of the local diet.

10) Southwest culinary culture area

The southwest culinary culture area covers Yunnan and large parts of Guizhou and Sichuan. As a mountainous region, this area produces rice, wheat, maize, sorghum and sweet potato, sources of the local everyday diet. There are not many food taboos in this area as there are no food prohibition rules set on pork, beef, mutton, chicken, duck and goose. Some ethnic groups have the custom of eating ants, locusts and tadpoles.

11) Tibetan Plateau culinary culture area

The Tibetan Plateau culinary culture area covers Qinghai-Tibetan Plateau and neighbouring provinces including Qinghai, Sichuan and Yunnan. The population of this area comprises mainly the Tibetan people. Farming and stock breeding are the two major modes of production of this area. Crops grown in the area include highland barley, barley and wheat. Maize and rice were also introduced into this area for cultivation. As for animal husbandry, cattle and sheep are the two largest groups of animal being raised. Beef, mutton, dairy products and wheaten food are part of the local diet with some items being more popular in some places. Air-dried food and raw food are consumed a lot while non-staple food such as vegetables is not available in large quantities.

1.2 Cooking Methods

Understanding how to use fire is a big leap forward made by the human beings, which made significant contributions to the development of food preparation process. The Chinese culinary culture has developed its own cooking methods over the years. Nearly 100 cooking techniques have been created including frying, stir-frying, sauté, deep frying, steaming and boiling. For everyday cuisine, roasting, boiling, steaming and stir-frying are the most commonly used methods.

1. Roasting

Roasting was probably the most primitive way of cooking. The primitive men accidentally ate roasted food made from natural open flame. After knowing how to use fire, they started to use fire to roast food.

Different techniques may be used for roasting process. According to Zhao Rongguang, "The primitive men placed large cuts of meat (mainly cervidae) on open flame allowing the fire to burn the meet directly. Or the meet was placed close to the fire without direct contact with the fire and the meat was cooked by using the heat. Or smaller cuts of food were cooked on a stone which had been heated at a very high temperature. For food unable to be cooked using the above three methods, the food may be packed in mud and put into fire or carbon ashes for cooking."[①] After invention of such cooking methods as steaming and boiling, roasting has remained an important way of cooking, mainly used for cooking meat. In some places, wheaten food is also made through roasting such as Nang of Xinjiang, baked pancake of Henan and thin pancake of Shandong.

Simmering is also one way of roasting. The food is placed deep in carbon ashes having just been burned and the food is cooked by using the afterheat of the carbon

---

① Zhao Rongguang. *Chinese Culinary Culture*[M]. Beijing: Zhonghua Book Company. 2012: 126.

Roasting

ashes. Many foods can be cooked through simmering process, mainly root vegetables. This technique has been used until today. Corn, potato, sweet potato and other foods can be cooked by using this method.

2. Boiling

Boiling probably originated during the early hunting period. Heated stones were placed in vessel containing water or placed in small puddle and food was cooked by using the heat of the stones. The real boiling technique was created after the invention of pottery ware. According to Zhao Rongguang, boiling is the primary technique having been used in China during the past 7,000-8,000 years for preparing dishes and staple food. If boiled in water, rice will become fluffy and the starch will dissolve in the water. Thus, nutrition is preserved and the boiled rice is flavorful and easy for digestion. Over the years, alternative techniques such as stewing and simmering have been developed through adjustment of heat and cooking time.

In today's China, boiling technique is adopted in every culinary culture area. Guangdong cuisine is famous for its stewed dishes. As the local saying goes, "We'd rather only have soup for a meal than have many dishes for a meal." Rich experience in stewing has helped Guangdong cuisine develop its unique style.

3. Steaming

Steaming is a cooking method where food is cooked by using the heat of steam. The ancient Chinese invented a utensil called "Zeng" (steamer) to steam food. At the 6,000 - 7,000 years old Hemudu cultural site, "Zeng" with one hole or several holes at its bottom was unearthed.

The cooking method can be used to steam rice including paddy rice and millet, or steam wheaten foods such as steamed bun, steamed stuffed bun, steamed twisted roll and Shaomai which are quite popular in today's China. Meat and fish can also be steamed such as steamed fish, steamed lamb, steamed duck, and steamed chicken. A famous dish of Hunan cuisine is steaming cured meat, cured chicken and cured fish together. Vegetables can also be steamed such as the dishes of spicy celery and steamed green beans from southern Shandong and steamed lotus roots from the south of the Yangtze River.

4. Frying

Frying is the most commonly used cooking method including simple stir-frying, stir-frying, quick-frying and special stir-frying. Frying was invented quite late. According to historical records, boiling was the primary cooking method during the pre-Qin period and the Qin & Han Dynasties. During the Wei, Jin, Southern and Northern Dynasties period, frying became popular. In real practice, a combination of cooking methods is usually used for food preparation rather than a single cooking method. For example, when preparing the dish of "Guota yellow croaker" from

Shandong cuisine, yellow croaker is first fried and then stewed. When making the dish of "fried potato, green pepper and eggplant" from northeastern Chinese cuisine, potato and eggplant are first deep-fried and then stir-fried.

1.3 Dietary Habits

1. Mealtimes

According to historical records, the Chinese people only had two meals a day, i.e., breakfast and dinner, during the pre-Qin period. According to oracle bone inscriptions, people of the Shang Dynasty divided a day into eight sections: morning (or "daybreak"), grand meal, Dacai, mid-day, noon, lesser meal, Xiaocai, evening. "Grand meal" and "lesser meal" refer to mealtimes. "Grand meal" is 7 o'clock to 9 o'clock and "lesser meal" is 15 o'clock to 17 o'clock. The practice of having two meals a day had been adopted for a long time and was still adopted during the Wei, Jin, Southern and Northern Dynasties period.

According to Zhao Rongguang, the practice of having three meals a day started to be adopted during the Spring and Autumn period. It was Zheng Xuan of the Eastern Han Dynasty who formally confirmed the practice of having three meals a day. When explaining the line of "having no meal at inappropriate time" from the chapter of "Xiangdang" of *the Analects of Confucius*, Zheng Xuan said, "Having three meals daily: in the morning, at noon and in the evening." This shows that having three meals a day started to become a common practice during the Han Dynasty. Today, having three meals a day is the predominant practice in China.

2. Dietary habits

Due to historical and geographical reasons, different region has developed its own dietary habits including arrangement of the daily three meals.

The north is known for its easy and quick breakfast. Typical food includes wheaten food and juicy food. For example, people in Beijing have sesame seek cake, steamed bun and steamed stuffed bun for breakfast. They also eat fried dough food such as deep-fried dough sticks, deep-fried dough cake and fried ring accompanied by beancurd jelly, fermented bean drink, wonton, mutton soup or different kinds of congee. Breakfast in Tianjin is similar to that in Beijing. Famous breakfast food items include Jianbing and seasoned millet mush. People in Shandong and Henan also have wheaten food and congee for breakfast. Shandong is famous for its Jianbing. Henan is known for its soup with pepper, the most famous version is from Xiaoyao township of Zhoukou city.

There is a wide choice of food for breakfast in the South, among which Canton morning tea is the most famous. Morning tea is what the Cantonese usually have for breakfast which offers different kinds of tea and snacks. The tea has gradually played a supporting role while the snacks have undergone improvement in both quality and variety. The staple food of morning tea is steamed stuffed bun. Representative steamed stuffed buns include steamed bun stuffed with barbecued pork, steamed dumpling stuffed with diced pig fat and sugar, small steamed bun stuffed with shrimp meat, small steamed bun stuffed with crab meat and all kinds of dry steamed Shaomai. Other snacks include flaky pastry, chicken congee, beef congee, sashimi congee, Chee Cheong Fun, shrimp fen, wonton and rice noodle roll. The most popular breakfast food items in Guangxi are all kinds of rice noodle roll and gruel of sweetened & fried flour. The locals of Changsha also love rice noodle roll and rice noodle roll restaurants are always packed every morning.

The eastern region abounds with rice and rice-based food is an integral part of the breakfast of the locals. For example, glutinous rice ball popular in Shanghai features rice ball stuffed with pickles (or sugar) and deep-fried dough sticks. Glutinous rice ball, deep-fried dough sticks, Chinese pancake and soybean milk are called the "Four Guardian Warriors" of a typical breakfast in the eastern region. Different from the north, noodles are popular breakfast food item in Jiangsu and Zhejiang such as seafood noodles of Zhoushan, knot noodles of Ningbo and stewed meat noodles of Suzhou.

With Qinling Mountains as the dividing line, the west is divided into two regions: the north and the south. As in Henan, breakfast food items in Shaanxi and Shanxi (two provinces in the northern region) include soup with pepper, sheep offal soup, Chinese hamburger and all kinds of congee. Breakfast food items of Chongqing and Sichuan which are situated in the South are usually spicy and flavorful. Representative items include spicy noodles, wonton, rice-flour noodles and hot & sour rice noodles.

Lunch and dinner are usually treated as major meals as they are the main source of energy required for daily work and learning activities. Compared with the fast and quick breakfast, lunch and dinner feature more items and larger portion of food. Due to the fast-paced lifestyle in big cities, people usually have a fast lunch. Despite the large portion of food eaten for lunch, the meal is often of poor quality making dinner the most important meal for the day. In medium-sized and small cities, especially in some cities in the North, there is a two-hour break at noon allowing the locals to have enough time to prepare a good square meal. As a result, people would have a quick and fast dinner in the evening.

A variety of food items are offered for lunch and dinner in each region including meat, vegetables, staple food and fruit. However, huge regional differences exist between the South and the North. People in the North like wheaten food and noodles are their favorite. Representative noodles include noodles served with gravy, hand-pulled noodles, stewed noodles and sliced noodles. In Zhengzhou, the locals would order a bowl of noodles at a stewed noodles restaurant or a hand-pulled noodles restaurant. Sometimes, the noodles are accompanied by one or two cold dishes. In the South, the locals would order a bowl of rice accompanied by one or two hot dishes, either meat or vegetables.

As for taste preference, there exist four groups: salty taste preference in the northern regions; sweet taste preference in the South and in the eastern region; sour taste preference in the northwest; spicy taste preference in the southwest. As people in the northern regions prefer salty taste, salt is the major flavoring used for cooking while sugar is seldom used. Sugar is widely used for cooking in the South and some people even add sugar for fried vegetables. Many northerners can't get used to the dishes after they come to the south for the first time. Shanxi is known for its vinegar so that people in the northwest like sour taste. Another explanation is that eating sour food may help prevent gall stones as the soil and water in the northwest is rich in calcium. In addition to people in the southwest such as Sichuan, Yunnan and Guizhou, people of Hunan and Jiangxi also love spicy food.

1.4 Cooking & Eating Utensils

Cooking & eating utensils are one of the key elements of culinary culture. Cooking skills are in part shaped by cooking utensils. For example, the invention of iron cooking utensils helped the development of the skill of stir-frying. In the course of a long history, China has developed a comprehensive system for cooking and eating utensils which includes pottery, porcelain, bamboo, wood and metal utensils.

Pottery cooking & eating utensils appeared early in human history. During the Paleolithic period and the Neolithic period, pottery ware had played a key role in everyday life. Pottery ware of different colors and usage also serves as important indicators for defining cultural characteristics of different historical period. Among pottery ware having been unearthed so far, most common objects include food container, container for storage, water container and boiler. In terms of appearance, there were kettle, bottle, rice steamer, urn, jar, stemmed bowl, wine vessel and cup. In addition to pottery ware, stone and jade utensils from the Paleolithic period and the Neolithic period were also found in some places.

Pottery ware remained as the commonly used cooking & eating utensils during the Xia, Shang and Zhou Dynasties with more beautiful appearance and patterns. Thanks to the development of bronze-casting techniques, bronze cooking & eating utensils had also been unearthed in large quantities. For example, jue (a kind of wine vessel) was discovered at archaeological sites of the Xia and Shang Dynasties period. Since the mid-Shang Dynasty, a wider variety of bronze ware had been invented including ding (cooking vessel), rice steamer, cooking tripod, plate and stemmed bowl. Early bronze ware pieces were roughly made with thin and light body. From the late Shang Dynasty to the Western Zhou Dynasty, the body became thick and the entire piece was more exquisite in appearance. Techniques for making jadeware and lacquerware (including bamboo ware and woodwork) had developed during the Spring and Autumn period as well as the Warring States period and elegant objects were produced during the two periods.

During the Qin and Han Dynasties, bronze ware pieces were no longer used as cooking & eating utensils. Glazed pottery, bamboo ware and woodwork became predominant as cooking & eating utensils. Bamboo ware and woodwork having been unearthed so far include container, box, wine vessel, kettle, cup, plate, ladle and spoon. Iron kitchen utensils started to be used during the Han Dynasty. Over 40 kitchen knives were found at the tomb of Nanyue King in Guangzhou. Wok became a commonly used object during this period.

The most significant change during the Wei, Jin, Southern and Northern Dynasties period was the wide use of porcelain objects. Compared with pottery ware, porcelain objects are more durable, beautiful and colorful. During the Eastern Han Dynasty, people already knew how to make porcelain objects. However, not until the Wei Jin period did porcelain objects begin to be widely used. During the Northern Song Dynasty, porcelain objects became the major cooking & eating utensils. The pot with green glaze and brown paint immortality picture unearthed at Dong Wu tomb in Changgang Village, Nanjing, Jiangsu Province, is the oldest under-glaze painted porcelain object discovered so far in China, which is of epoch-making significance in the history of porcelain development.

Since the Sui and Tang Dynasties, materials used for making cooking & eating utensils have not changed while appearance of the utensils has undergone significant changes. With increased cross-border exchanges, utensils of Persian, Qidan, Jurchen and the Western Xia styles were found on the dining table of Chinese people. Precious metal utensils such as porcelain utensils, glass utensils and gold/silver utensils also take on a variety of shapes.

Cooking & eating utensils used by Chinese people today are mostly made of iron and porcelain. They can be classified into four categories: cooking utensils; tableware; tea set and drinking vessels.

Usually made of metals, cooking utensils include wok pan, frying-pan, steamer and saucepan. Most wok pans and frying-pans are made of iron. Containing no other chemical substances, iron pan does not oxidize. No dissolved substances break off when using iron pan for frying or boiling food. Even there are iron substance dissolved, they are good for health. In addition to iron steamer, there are aluminum steamers. Iron pan is apt to rust, particularly when coming in contact with water. Aluminum pan is light, durable, ready for heat conduction and not apt to rust. It's ideal for steaming and boiling food. However, aluminum may dissolve at high temperature posing a health risk. Most saucepans are casserole. With good aeration and absorbability, casserole is suitable for stewing food as heat can be distributed evenly and dissipated slowly.

Tableware are used for dining. They include bowls, chopsticks and spoons. Tableware can be made of metal, ceramic, glass, paper, plastics and bamboo wood. Most tableware is ceramic items.

Tea set is a collection of items used for drinking tea. The basic components of a tea set are teapot and cups. Auxiliary items include tray, bowls and teaspoon.

Drinking vessels are used for drinking wine. Commonly used items include wine pot and wine glass. Most contemporary drinking vessels are made of ceramics and glass. Some are made of metal, jade and bamboo wood. On the basis of the shape, drinking vessels have been given different names including zun, hu, qu, min, jian, hu, gong and weng. Contemporary wine glasses can be divided into three groups: (1) White wine glass which is small in size and has a capacity of about 30 milliliters. Most white wine glasses are ceramic items or made of glass; (2) Red wine glass usually has a capacity of around 300 milliliters. Introduced from the West, most red wine glasses are goblets made of glass; (3) Beer glass usually has a large volume and its capacity may reach 500 milliliters. Most beer glasses are made of glass.

Chapter 2 Cuisine for Festivals

Chinese civilization has existed for 5,000 years. In the course of the long history, the Chinese people have established many seasonal festivals and developed their own folk customs. These festivals bring happiness to the Chinese people and demonstrate the harmonious relationship between nature and the people as well as the strong connections among the people. As for festival celebrations, festival food is one of the most interesting elements. Each festival has its own representative foods and each food is associated with the old traditions and carries a deep meaning.

Foods for Chinese seasonal festivals can be classified into three categories:

(1) Offerings for sacrificial ceremony. In ancient times, offerings played a central role in such ceremonies as sacrificial rites and celebrations which were held by the imperial palace, the local governments, the patriarchal clans or the households. In today's China, most places no longer engage in these activities. Only similar activities of symbolic significance are now held in a few remote areas or at some special occasions.

(2) Food to be eaten for festival celebration. Throughout the year, the Chinese people would celebrate a number of seasonal festivals. Specific food items are offered for each festival in order to follow the folk customs. For example, on the eve of the Lunar New Year, every household in the northern regions eat jiaozi to express the wishes for family reunion and happiness. Eating zongzi (sticky rice dumplings) during the Dragon Boat Festival is a long standing tradition to commemorate Qu Yuan, the patriotic poet, and celebrate the memory of the country's past. Eating moon cakes at Mid-Autumn

Festival reflects people's wishes for family reunion as the round moon cake is a symbol of the full moon.

(3) Gifts given to friends, relatives or other people for festival celebration or other special occasions. In Chinese culture, giving gifts is a way of expressing friendliness and maintaining interpersonal relationships. For example, people offer moon cakes at Mid-Autumn Festival and zongzi at Dragon Boat Festival as gifts for relatives and friends, as the folk customs regulate. During the Chinese New Year, people also bring gifts for their friends and relatives when they pay a New Year call.

2.1 Spring Festival

The name of "Spring Festival" was created in 1911, but the festival itself had 3,000-plus years of history. According to the lunar calendar, the first day of the first lunar month is called the "first day of the year". In ancient times, the day was called the "New Year's Day". Celebration of the Spring Festival, lasting for about one month, starts on the 23rd day of the twelfth month of the preceding lunar year and ends on the 15th day of the first month of the lunar year (which is also called the Lantern Festival). The final month of a lunar year is called the "twelfth month of the lunar year". In ancient times, sacrificial activities were held during this month in honor of the gods and the ancestors. Starting from the 23rd day of the twelfth month of the lunar year, people begin to prepare for the celebration of the Spring Festival. On the 23rd day or the 24th day of the twelfth month, people would engage in "kitchen god worshiping" activity. Kitchen god is the most important god of every household as he watches closely what is happening on the earth. He returns to heaven to report to the Jade Emperor① about the deeds of the household. Among the food items offered to kitchen god, there is a candy called "sweet melon". People

① Jade Emperor: The greastest god in heaven.

hope to seal the lips of kitchen god with the candy so that he would not report about their misdeeds to the Jade Emperor. Other customs during the Spring Festival include hanging peach wood charms, pasting door-god, pasting Spring Festival couplets, hanging New Year pictures and setting off firecrackers and fireworks.

The last day of the preceding lunar year is called the "New Year's Eve". On New Year's Eve, all the family members gather for a reunion dinner. The reunion dinner is of extraordinary significance as it signifies the coming of the New Year as well as family reunion, harmony and happiness in the coming year. No matter how far away or how busy they are, people would return home to celebrate the festival with their parents and enjoy the reunion dinner together. A reunion dinner is usually given on the 29th day (if the month only has 29 days according to lunar calendar) or 30th day of the last month of the preceding lunar year. The New Year's Eve dinner is called "Spring Festival reunion" or "family reunion". By having the dinner, people bid farewell to the old year and usher in a new year and therefore, the New Year's Eve dinner is also called "dividing the year". In ancient times, the New Year's Eve dinner was regarded as a way of curing disease, warding off the evil spirit and enhancing health. People express their hope for the coming year through each dish of the dinner. For example, a simmering and enticing hot pot represents prosperity; a dish of fish signifies "surplus and good luck" and "abundance for every year" as the word for "fish" is a homonym of the word for "surplus"; radish is also called "caitou" in Chinese which is a homonym of the word for "good luck"; fried food such as shrimp and spring roll carries the meaning of family prosperity; desserts stand for wishes for a happy life in the coming years.

During the Spring Festival, friends and relatives often dine together and each meal may have wine and meat. For such occasion, people may prefer low-fat food over greasy food. Therefore, the menu for a Spring Festival banquet must have both meat dishes and vegetable dishes. For a meal, there should be cold dishes, hot dishes and snacks. As low-fat cold dishes are refreshing, they can accompany the wine. As for hot dishes, they should contain both tender food items and crispy food items such as fried food. The dishes may be red, white, green and yellow in color so as to reflect the aspiration for a happy new year. If the diners want to eat with great relish, the dinner should have sweet, salty, sour and spicy dishes.

There is also a sequence for serving the dishes. As summarized in the past, "salty food served first, light food comes later; thick dish comes first and followed by clear food; dish without soup served first and dish with soup comes later." Cold dishes are usually served first and hot dishes are served later. As for stir-fried dishes, salty dishes are served first as salty food may serve as an appetizer. Fried food and desserts may be served between the other courses or served before the main course since the greasy fried food and desserts can give the diners a feeling of satiety. Sour and light dishes with soup are served last as they may dispel the effects of alcohol and reduce the greasy feeling. Snacks are served after meal in order to add delight to diners' gustatory experience. Contemporary people attach more importance to healthy diet and balance of nutrition. Instead of eating and drinking too much during the Spring Festival, people pay more attention to the intake of vegetables and fruits so as to achieve nutritional balance as well as acid-base balance of metabolites, and avoid consequences such as hyperlipemia and sudden gain in weight.

Except delicacies, wine is a must-have at a banquet of the Spring Festival. However, as those gathering for reunion dinner include both the young and the old who may have different drinking habits, white spirit, wine and hot drinks for non-drinkers such as fresh juice, corn juice and milk can be prepared in advance.

Here are some commonly seen dishes at a Spring Festival banquet:

Cold dish: sweet and happy (honeydew red dates); garden overflowing with spring vegetation (chrysanthemum with old vinegar); abundant wealth (ham and cabbage)

Main course: surplus for every year (steamed sea grouper); good luck (steamed fish head with diced hot red peppers); good luck for the coming year (iron platter calf ribs); happy reunion (scaled shrimps)

Hot dish: family reunion (Chinese flowering cabbage with mutton balls); prosperity (fermented soybeans with peppers); honor and wealth (broccoli with squid); abundant wealth and many children in the family (pine nuts and corn)

Soup stew: snow pear and banana stew; catfish and eggplant stew; lettuce and dried shrimps soup; white corn soup

Staple dish: jiaozi (popular in most places in the North; the dish of "jiaozi with egg and Chinese chives stuffing" meaning acquiring wealth); nian gao (popular in most places in the South); getting promoted step by step (multi-layer steamed bread with cheese); pumpkin pie; family happiness (Yangzhou-style fried rice)

The reunion dinner on the New Year's Eve usually has a number of courses including chicken, duck, fish and meat. Chicken represents good luck and fish stands for surplus for every year. In some riverside areas and coastal areas such as Shanghai and Ningbo, seafood is also found on the dining table of the reunion dinner. On the New Year's Eve, the Chinese people follow the custom of staying up late or all night. Family members gather for the reunion dinner drinking, chatting or playing games. As a way of wishing the children health and good luck in the coming year, the elder would give money to the children as a gift. The money is then placed under the children's pillow.

Food items related to the Spring Festival include Laba congee, nian gao, jiaozi and glutinous rice ball.

1) Laba congee

Chinese people eat "Laba congee" at "Laba Festival". "Laba Festival" falls on the eighth day of the twelfth lunar month when the ancient people worshiped their ancestors and gods, and prayed for harvest and good luck. Some places have the custom of eating Laba congee on this day. According to historical records, the tradition of eating Laba congee began during the Song Dynasty and lasted over 1,000 years. On the day of the festival, every household, rich or poor, eat Laba congee. The earlier version of Laba congee was made of boiled red beans. Then, different regions developed their own distinctive version and added more ingredients to the congee. Laba congee is both a seasonal delicacy and a dish for health preservation. If eaten during the cold weather, it can help nourish the spleen and stomach. As time goes by, different versions of Laba congee have been developed. Currently, Laba congee becomes a snack with distinctive local flavor in many places.

Laba congee

2) Nian gao

The Chinese people have the tradition of eating nian gao during the Spring Festival. Nian gao has the symbolism of "raising oneself higher in each coming year." Eating niao gao during the Spring Festival, therefore, represents the wishes for more successes at workplace and a happier life in the coming year. Nian gao has different versions: white and yellow rice cake in the North; nian gao made from finely ground rice flour in regions south of the Yangtze River; red turtle cake in Taiwan. Nian gao is made of steamed rice flour through such techniques as kneading and smashing. In regions south of the Yangtze River, local people usually add water to sticky rice and grind it into rice pulp which is then steamed and made into bar-shaped or brick-shaped nian gao.

Legend has it that the tradition of eating nian gao originated in the capital of Wu state (today's Suzhou of Jiangsu Province) during the Spring and Autumn period and the Warring States period, and then the other places in China started to adopt the tradition. There is a folk proverb in Ningbo which says, "nian gao takes you higher year after year, living a happier life this year than last year." Printing plate is often used to press nian gao into the shapes of "five blessings", "six treasures", "money" and "ruyi" which symbolize "good luck" and "favorable auspices". Some nian gao are made into the form of "jade hare" or "white goose", a perfect combination of content and form.

3) Jiaozi

Jiaozi came from *jiaozi* in ancient times. Originally called *jiao'er* (tender ears), jiaozi were invented by Zhang Zhongjing, known as a sage of Chinese medicine in China, and had a history of more than 1,800 years. Legend has it that, in order to help the poor fight frostbite, Zhang Zhongjing filled dough wrappers with cold dispelling medicines (mutton and black pepper etc.) so as to treat disease. Fillings of jiaozi at that time were medicine used to treat disease. Jiaozi is a tradition food item popular among the Chinese people. It's the staple food and snack in the north China as well as a popular dish for festival celebration. As Jiaozi is a symbol of wealth, silver or shoe-shaped gold ingot is sometimes hidden inside jiaozi, a practice representing Chinese people's aspiration for wealth. On New Year's Eve when all the family members sit around the table making jiaozi, the scene symbolizes that they are working together to make "shoe-shaped gold ingot". At midnight, jiaozi are poured into the wok to be boiled. This is the first meal of the

New Year through which people express their wishes for happiness in the coming year.

Cold water and flour are used to make jiaozi. Mix water and flour together and knead into a big dough. Use knife to cut the big dough into several small doughs or simply use hand. Use a small rolling pin to roll out these small doughs into wrappers whose center is thicker than the side. Fill the wrapper with stuffing and pinch the wrapper till it is in a crescent shape or horn shape. Wrapper can also be made from dough made with boiling water, pastry or rice flour. Meat or vegetables can be used as fillings which can be sweet or salty. Methods of cooking jiaozi include: boiling, steaming, flipping, frying and deep frying etc. Meat fillings include three delicacies, shrimp meat, crab roe, sea cucumber, fish, chicken, pork, beef and mutton etc. Vegetable fillings include mixed vegetables and ordinary vegetables. Jiaozi features with thin wrapper and tender fillings. With flavorful taste and distinctive shape, jiaozi is a delicacy you'll never get tired of eating.

Jiaozi

2.2 Lantern Festival

The traditional Lantern Festival falls on the 15th day of the first lunar month. It is called the Spring Lantern Festival and the celebration is called the "happiness of the Lantern Festival". On the evening of the festival, lanterns are hanged everywhere. People appreciate the beauty of the lanterns, solve riddles on the lanterns, watch performance and have fun.

There are many tales about the origin of the Lantern Festival. Generally, the origin is associated with religious activities. As some believe, Buddhism was introduced into China during the Han Dynasty, and the emperor gave an order to light lanterns on the 15th day of the first lunar month to show respect for the Buddha. Another tale about the origin of the festival is associated with Taoism. As some say, the birthday of a god of Taosim fell on the 15th day of the first lunar month. People lighted lanterns as a way of worshiping the god in hope of obtaining protection and care from the god. Still, some believe that the origin of the Lantern Festival had something to do with the worship of fire by the primitive people.

The food for the Lantern Festival is yuanxiao, originally named tang yuan. Eating tang yuan at the festival symbolizes "family togetherness similar to the full moon." As the festival is on the day when there is the first full moon of the new lunar year. Eating tang yuan on this day expresses people's wishes for family union and happiness. As a food item, "yuanxiao" has a long history in China. During the Song Dynasty, there was a popular food eaten at the Lantern Festival which was called "floating yuanzi" and later called "yuanxiao". The business people also named it "shoe-shaped gold ingot". Yuanxiao is called "tang yuan" in southern China. The fillings are usually made from white sugar, rose, sesame seeds, sweetened bean paste, osmanthus, walnut meat, nuts and jujube paste. Glutinous rice flour is made into round shape filled with meat or vegetable fillings. With a number of variations, cooking method of tang yuan also varies including boiling, frying and steaming. Eating tang yuan symbolizes family reunion and happiness. Instead of filling dough with stuffing, tang yuan in Shaanxi is made by "rolling" the stuffing in glutinous rice flour. And tang yuan is cooked through boiling or frying.

In addition to yuanxiao, each region has its own distinctive food items for the Lantern Festival. For example, people in the northeast eat frozen fruit and frozen fish during the festival. People in Guangdong have the custom of "stealthily" picking romaine lettuce during the festival. The romaine lettuce is then boiled with cakes. The dish symbolizes wishes for good fortune.

Yuanxiao

2.3 Cold Food Festival

In ancient China, a major festival named Cold Food Festival was observed

105 days after the Winter Solstice and one to two days prior to the Qingming Festival. When the Cold Food Festival was first inaugurated, people were prohibited from lighting fire and were only allowed to eat cold food as a way of observance. Other customs had been incorporated into the celebration over the course of time including visiting and tidying up tombs, having an outing in spring, playing on the swing, cuju[①] and cockfighting. The Cold Food Festival had been observed in China for over 2,000 years and was once dubbed the most important memorial day for the common people. Among all the traditional festivals, the Cold Food Festival is the only one whose title was named after a culinary custom.

Legend has it that the Cold Food Festival was created to commemorate Jie Zitui, a loyal subject during the Spring and Autumn period who was burned to death in the mountains. Jie Zitui was a native of Jiexiu in today's Shanxi. He was a virtuous official of the the State of Jin and was respectfully called Jiezi by later generations. Reluctant to serve in the same administration along with those small men who were eager to curry favor with those in power, Jie Zitui retired to hermitage in the mountain with his mother. Having failed to entice Jie Zitui out of the mountain, Due Wen of the State of Jin gave an order to set fire to the mountain with the aim to smoke Jie Zitui out of hiding. Unexpected, Jie Zitui, with his mother in his arms, was burned alive under a large willow tree. To remember this virtuous official, Due Wen issued a decree forbidding people to light a fire for cooking on the anniversary of Jie Zitui' death. Only cold food was allowed for consumption on this day, thus the creation of the Cold Food Festival. As every household was prohibited from lighting a fire for cooking and only cold food was allowed to be eaten during the festival, the food was apt to deteriorate harming people's health. Therefore, the Cold Food Festival and the Tomb-Sweeping Festival were observed as one festival after the Tang Dynasty, given the fact that the two festivals were originally only days apart. Since then, the Tomb-Sweeping Festival

---

① "Cuju" refers to the activity of treading or kicking ball in ancient China, similar to today's soccer.

gradually became a major festival for commemoration and veneration of ancestors.

Food items eaten during the Cold Food Festival include: (1) Hanshi congee. Since no household was allowed to light a fire during the festival, they would prepare congee made from barley in advance which was ready to be eaten during the festival, thus the name "Hanshi congee". (2) Runbing vegetable or tender shoot vegetable. A variant of spring roll, runbing vegetable originated in Quanzhou and became popular in Taiwan, Fujian and other places. (3) Black rice. One of the major food items for the Cold Food Festival, black rice is made by dyeing glutinous rice with juice of black rice tree and then boiling the black colored rice. (4) Happy balls. In Chengdu of Sichuan, fried rice is made into balls which are linked together with thread. The balls are either small or big and are dyed with

Hanshi congee

different colors, thus the name "happy balls". In the old times, the food was sold along the road from the outside North Gate of Chengdu to "Huanxi Nunnery" (Huanxi means happy). Scores of drinks are consumed during the Cold Food Festival including wine brewed in spring, new tea and clear spring water. Food eaten during the festival varies regionally. People in southern Shanxi eat bean jelly, cold noodles and cold cake. People in northern Shanxi eat *chaoqi* (cut steamed cake flour or white flour into squares the size of dice, dry them in the sun and then fry them with soil till they become yellow) during the festival. In some mountainous regions, the entire family eats fried noodles (mix fried coarse cereals with all kinds of dried preserved fruits and grind them into flour).

According to the custom, "cold swallows" are to be steamed to observe the Cold Food Festival. Flour is made into thumb-sized swallows, birds, animals, fruits and flowers which are then steamed and dyed. They then are stuck on the needles of wild jujube tree as a decoration for the room or as a gift sent to other people.

2.4 Qingming Festival

The Qingming Festival (Tomb-Sweeping Festival) is the only Chinese traditional festival which is named after a solar term. Created during the Zhou Dynasty, the festival has a history of more than 2,500 years. Over the centuries, the Qingming Festival has carried a deeper meaning and different regions have developed their own customs. Nevertheless, sweeping the tombs, worshiping the ancestors and going for an outing remain the primary themes of the festival. By sweeping the tombs of their ancestors and remembering their deceased, Chinese people express their deep love for their family. On the day of the festival, people go for an outing enjoying the beautiful scene of the spring.

There are a lot of culinary customs for the festival in China. In regions south of the Yangtze River, green dumplings are eaten during the festival. When making green dumplings, juice of brome grass and glutinous rice are mixed and made into doughs. Sweetened bean paste and jujube paste are then used as stuffing. The doughs filled with stuffing are then placed on steamers with common reed leaves underneath. The steamed doughs are green in color and release aroma. Green dumpling is the most distinctive food in the region for observance of the Qingming Festival as well as an indispensable item for worshiping the ancestors.

Green dumplings

During the Qingming Festival, every household in Huzhou of Zhejiang makes zongzi which can serve as offerings for tomb sweeping and food for outing. As the saying goes, "Qingming zongzi is solid." Around the Qingming Festival, snails get fatty and each rural household would have snails as a delicacy. Snail meat is picked with a needle for cooking, a process called "picking the green". Afterwards, snail shells are tossed on the roof and the noise made is believed to be able to scare off the rats. There is a saying in Heshan Township, Tongxiang of Zhejiang, which goes, "The Qingming Festival is as important as the Lunar New Year." On the evening of the Qingming Festival, the entire family gathers for a reunion dinner. A number of dishes would be an indispensable part of the menu: stir-fried snails, lotus root stuffed with glutinous rice, sprouted broad beans and Indian Kalimeris Herb. Eating lotus root is a way of expressing hope for long and high-quality silk produced by silkworms. Eating sprouted beans is a way of expressing wish for prosperity as the Chinese word for "sprouted" is the same for the Chinese word meaning "prosper".

People in China also eat fried dough twist during the Qingming Festive. A fried food item, fried dough twist is crispy and attractive in appearance. It was called "cold item" in ancient times. The southern regions and the northern regions make different style of fried dough twist: fried dough twist of the northern regions is large in size and made from wheat flour; fried dough twist of the southern regions is delicate in appearance and made from rice flour. In areas inhabited by the ethnic minority groups, a variety of fried dough twist is made with different flavors. Fried dough twist made by the Uygur ethnic group, the Dongxiang ethnic group, the Naxi ethnic group and the Hui ethnic group are the most famous.

During the Qingming Festival, nutritious food items are also consumed in different places such as chicken, cake, Qingming zongzi, glutinous rice cake, Qingming sticky rice cake, and dried congee. As people's living standards have improved, pot-stewed meat and salty tea egg are also consumed during the holiday.

2.5 Dragon Boat Festival

The Dragon Boat Festival, or "The Duanwu Festival", is the second largest traditional holiday in China. Also known as "Duanyang Festival" and "Fifth Month Festival", the Dragon Boat Festival falls on the fifth day of the fifth lunar month. It's a holiday with over 2,000 years of history. "Duan" means "the beginning" and "Wu" is the homophone of "five" in Chinese, and therefore "Duanwu" actually means "the fifth day".

There are a lot of tales about the origin of the festival. The most widely known is the story about Qu Yuan, a great poet during the Spring and Autumn period and an official of the State of Chu. Since the king of the State of Chu refused to listen to sincere advice and trust in small men instead, the State of Chu was finally defeated by invaders. Qu Yuan committed suicide by drowning himself in the Miluo River with a large rock in his arms, died for his country. In great grief, the local people went to the Miluo River to mourn for him. Some rowed boats on the river trying to retrieve his body. Some dropped rice balls, eggs and zongzi into the river so that the fish would eat them instead of Qu Yuan's body. Some poured realgar wine into the river hoping the medicinal wine may get the dragons and water beasts dizzy so as to prevent them from harming Qu Yuan. Ever since, the tradition of eating zongzi and drinking realgar wine on the fifth day of the fifth lunar month was created in commemoration of Qu Yuan, the patriotic poet.

In addition to eating zongzi, a shared tradition in China, each regions has its own culinary traditions for the Dragon Boat Festival. For example, people in the Jianghan Plain eat ricefield eel; people in Gansu eat fan-shaped cake; people in Nanchang of Jiangxi eat tea egg; rural people in Henan and Zhejiang eat garlic egg; people in Yanbian of Jilin eat glutinous rice cake; people in Jinjiang of Fujian eat fried glutinous rice balls with sesame; people in Wenzhou eat thin pancake; people in Huili of Sichuan eat stewed root.

Ancient Chinese people attach great importance to the Summer Solstice and the Winter Solstice, two solar terms marking the onset of a new season. The Summer Solstice had been an important festival which was superseded by the Dragon Festival. The Summer Solstice usually occurs in mid-June on the Gregorian calendar. After the Summer Solstice, the night becomes longer and the day is shorter. During this period of the year, the plague of floods and pests often caused harm to people's health. As the ancient Chinese people believed, the masculine energy was reduced and the feminine energy was increased in May. Various kinds of "evil" such as pest, disease would appear during this month. Therefore, May was regarded as an "evil month" by the common people and the fifth day of the fifth lunar month was viewed as the most unpropitious day of the year. As a result, people would hang mugwort and calamus above windows during the Dragon Boat Festival. Mugwort is an aromatic traditional Chinese medicine which is able to dissolve turbidity and dispel plagues with great effects. It's also called "mugwort tiger" by the ancient people. Calamus is believed to be able to prevent diseases and remove evil spirits. If hung under the eaves, it's called "calamus leaves". In the past, many households would cut colored paper into the patterns of "elimination of five poisons" or auspicious gourd curtain flowers. Girls would use colored thread and soft silk to make delicate five-colored zongzi, tiger and head of garlic. These objects would then be strung together and hung on hairpins or the front of the clothes of children.

The main food item eaten during the Dragon Boat Festival is zongzi. On the day of the festival, people also drink realgar wine.

1) Zongzi

Also known as "jiao shu" (triangular lump of glutinous rice wrapped in a leaf) and "tong zong" (tube-shaped zongzi), zongzi are made by wrapping glutinous rice and fillings with bamboo leaves and cooked by boiling. Made of glutinous rice, zongzi is wrapped in bamboo leaf and made into a triangular shape. Over the centuries, more variants are added to the zongzi family. In addition to ox horn zongzi, there are conical zongzi, weight zongzi, water chestnut-shaped zongzi, cloth-wrapper shaped zongzi, pillow zongzi and even pavilion-shaped zongzi as made in the imperial palace in ancient times. To celebrate the Dragon Boat Festival, every household would soak glutinous rice in water, wash bamboo leaves and wrap zongzi. An increasing number of varieties of zongzi have been created over the years. In terms of fillings, red jujube and mixed beans are often used as stuffing in the northern regions. Other types of zongzi made by people in the northern regions include: nut fruit zongzi with peanuts, pine nuts and other nut fruit as fillings; preserves zongzi with preserves as fillings; polished glutinous rice zongzi; red beans and jujube paste zongzi; wheat zongzi; walnuts and pine nuts zongzi. Fillings used in the southern regions include sweetened bean paste, meat, eight treasures, ham and yolk. Other representative zongzi of the southern regions are: ham and smoked meat zongzi; dried scallops and shrimp meat zongzi; Fujian and Taiwan carbonado zongzi.

Zongzi

2) Realgar wine

The day of Duanwu was viewed as an evil day by the ancient people. Therefore, drinking realgar wine on this day became one of the most important culinary traditions of the Dragon Boat Festival, especially in Yangtze River Basin area. As the ancient saying goes, "all the diseases will be prevented once you drink realgar wine." This shows the custom of drinking realgar wine has become an important way for Chinese people to pray for good fortune, dispel evil spirits and avoid disasters.

Realgar is both a traditional medicinal material and a mineral substance. Commonly known as "arsenic disulfide", the primary content of realgar is arsenic sulfide. It also contains mercury which is poisonous. Realgar wine is made by adding a small amount of realgar into white spirit or home-made yellow rice wine. Wine made purely from realgar can't be used an alcoholic drink. Effects of realgar wine include disinfection, expelling insects and detoxicating five poisons. The five poisons are snake, scorpion, centipede, gecko and toad. Drinking realgar wine can help kill the "five poisons". In traditional Chinese medicine, realgar is also used to treat skin disease. During the ancient times when there was no disinfectant such as iodine, realgar wine was used to remove toxin and relieve itching. Therefore, drinking realgar wine is beneficial for health and scientifically reasonable. The adults also use realgar to draw a dot or the Chinese character of "Wang" on the forehead of the children, a method for disinfection and protection against evil spirit as well as a way of expressing the adults' wishes for health and good fortune for the children.

Celebration of the Dragon Boat Festival is not as grand as that of the Spring Festival or the Mid-Autumn Festival in terms of dishes served. However, every household would prepare some delicacies for the seasonal festival. The banquet held for the Dragon Boat Festival is usually in the form of family feast which is not sumptuous. There are a variety of seasonal vegetables available around the time of the Summer Solstice such as edible gourd, green garlic, cucumber and fresh corn. As for aquatic products, yellow croaker and ricefield reel are in season. People can choose whatever they want.

As for dim sum, in addition to zongzi, "five poisons cake" (pastry with pattern of five poisons) and "rose cake" are also made for the celebration. During the Dragon Boat Festival, cherry, mulberry and loquat are available which were seasonal fruits prepared for celebration of the festival in the past.

As for beverages, "calamus wine" is prepared for celebration in addition to beer and low strength spirit. Made from traditional Chinese medicine, calamus wine used to be a must-have for celebration of the Dragon Boat Festival. Nowadays, it's available almost year-round on the market, ready to be consumed during the Dragon Boat Festival. If calamus wine is not available, mulberry wine can be used as a substitute. As a seasonal wine for the Dragon Boat Festival, mulberry wine has the effects of "nourishing the five internal organs and improving hearing & eyesight".

2.6 Mid-Autumn Festival

The Mid-Autumn Festival falls on the fifth day of the eighth lunar month. Also known as "Festival of the Eighth Month" and "Reunion Festival", it's a traditional festival celebrated by many ethnic groups in China. It's said that the moon is closest to the sun and the moon is the biggest, the roundest and the brightest on this day. Consequently, there has been a custom of having a banquet and gazing at the moon on the day of the festival since the ancient times.

The Mid-Autumn Festival is also celebrated by over 20 ethnic minority groups in China. The Dong ethnic group calls the festival "pumpkin festival" while Mulao ethnic groups calls the festival "Youth Festival", slightly different from the Han people's version.

The origin of the Mid-Autumn Festival is associated with the worship of the sun and the moon by the ancient Chinese people. The Mid-Autumn Festival was formally celebrated during the Song Dynasty. During the Ming and Qing Dynasties, the Mid-Autumn Festival became a traditional festival which was as important as the Spring Festival. Activities conducted during the festival included worshiping the moon and praying for harvest. During the Ming Dynasty, people in Beijing would have the ceremony of worshiping the moon. Round fruits and cakes were offered as sacrifices to the god of moon. The god of moon which is called "the rabbit master" of the

Beijing folk custom came into being during that period.

Moon cake is the representative food for Mid-Autumn Festival celebration. As for celebration of other festivals, food items of local distinctive features are also consumed during the Mid-Autumn Festival across the country. For example, people in Beijing eat wine preserved crab, people in Fujian eat areca taro roast duck, people in Xixiang County of Shaanxi eat watermelon cut into the shape of lotus flower, people in Shanghai drink osmanthus-scented honey wine while eating, people in Nanjing eat salted duck, people in Hangzhou eat bass stewed with water shield, people in Sichuan eat smoked duck, sesame cake and honey cake, and people in Guangdong eat taro and glutinous rice cake.

Moon cake is the No. 1 delicacy for Mid-Autumn Festival celebration. The round moon cake symbolizes "reunion". Also called "*hu* cake", "palace cake", "small cake", "moon dumpling" and "reunion cake", moon cakes used to be the offerings given to the god of moon when the ancient people celebrated the Mid-Autumn Festival. Moon cake gradually became an indispensable part of Mid-Autumn Festival celebration.

The term of "moon cake" first appeared in the writings from the Southern Song Dynasty. In Chapter Six titled "Steamed Foods for Distribution" of Zhou Mi's *Reminiscence of Wulin*, a number of steamed food items are mentioned including "lotus leaf shaped pancake", "lotus pancake", "mutton steamed buns", "vegetable cake" and "moon cake." In Chapter Sixteen titled "Meat and Vegetables Shop" of Wu Zimu's *Meng Lianglu*, food items mentioned include "chrysanthemum cake", "moon cake" and "plum cake." In these two books, "moon cake" has no association with the Mid-Autumn Festival. As *Shan Fulu* mentions "moon soup" instead of moon cake when talking about "seasonal food" of the Mid-Autumn Festival and no writings from the Yuan Dynasty mentioned moon cake, the "moon cake" mentioned by the people of the Song Dynasty is probably an ordinary moon-shaped food item rather than the moon cake which is an indispensable part of the Mid-Autumn Festival celebration.

Moon cake used for Mid-Autumn Festival celebration is formally mentioned in the writings from the Ming Dynasty. For example, in Volume Two titled "Happy Memories of the Prosperous Dynasty" of Tian Rucheng's *Tour of the West Lake*, it says "The Mid-Autumn Festival falls on the 15th day of the eighth lunar month. People give moon cakes to each other as a gift as moon cake symbolizes reunion." In the section titled "Sending Moon Cakes as Gift during the Eighth Lunar Month" of Shen Bang's *Essays on Wan Government Office*, it says, "Both the scholars and the common people give moon cakes to each other as a gift. The moon cakes vary in size. The shops make different kinds of moon cake by using fruits as fillings. One type of moon cake even costs 100 pennies." According to these writings, it was during the Ming Dynasty that moon cake started to become an indispensable food item for Mid-Autumn Celebrations. Therefore, giving moon cakes to each other as a gift during the Mid-Autumn Festival has become an important custom since the middle of the Ming era. Different styles of moon cake with distinctive regional flavor have been developed over the years including Beijing-style moon cake, Cantonese-style moon cake, Suzhou-style moon cake, Teochow-style moon cake. New styles of moon cake have been created over time and the techniques of making moon cake have also improved. Now, there are salty, sweet, meat and vegetables moon cakes. Some moon cakes are plain in appearance while some moon cakes are adorned with decorative edge. Different region has its own unique style of moon cake. For example, Cantonese-style moon cake is featured with thin skin, soft texture, sweet taste and delicious fillings. Suzhou-style moon cake is crispy, tasty and not excessive in sweet or salty taste. Teochow-style moon cake uses sweetened white gourd as fillings and offers soft and crispy texture.

The Mid-Autumn Festival is an occasion for family reunion. Every household prepares a sumptuous dinner for the celebration. The menu of the dinner depends on the culinary habits of the host and the guests as well as the habits of the local people. Different region prepares different food items. As for traditional food, there are stewed fillings, leek, shaomai, roast pig, south oven duck, taro, crab and

Moon cake

bream fish etc. Seasonal fruits include grapes and binzi. After the solar term of White Dew, chestnuts come into season which can be used for cooking dishes such as "stewed chicken with five fruits". As for alcoholic drinks, the crystal clear osmanthus wine is an ideal choice considering its rich aroma of osmanthus and grape. Osmanthus wine made in Beijing has won several state-level awards for its high-quality and unique taste.

Cold dishes for the Mid-Autumn Festival banquet include shredded ginger with tripe, tricolor cake, sweet and sour hairtail, green zizania latifolia with dried sea shrimps and edible rape. Hot dishes include fried pork liver with scallions, shrimp balls, crab stewed with tendons, white jade jiaozi, roast duck with chestnuts and steamed bream fish. Typical soup dish includes stewed chicken with five fruits. Staple food includes fried rice with osmanthus, which is a seasonal dish.

On the evening of the festival, people gather together, either at a restaurant or in the courtyard of their home, gazing at the orange osmanthus in the moon, appreciating the fragrant osmanthus and drinking osmanthus-scented honey wine. It's really a great time for family reunion and festival celebration.

2.7 Double Ninth Festival

The Double Ninth Festival (Chongyang Festival), a traditional Chinese festival, falls on the ninth day of the ninth lunar month. To celebrate the festival, the Chinese people have the custom of climbing mountains, wearing *zhuyu*[①] plant and drinking chrysanthemum wine. In Chinese culture, the odd number is yang number and therefore nine is a yang number. The ninth day of the ninth lunar month has two numbers of "nine", hence the name "double nine". Traditionally, the common Chinese people believe that this day is an ominous day. To protect against the danger[②] of this day, people would "climb mountains" and "wear *zhuyu* plant". On the day of the festival, people in some places would drive their farm livestock away from home. Legend has it that the activity of climbing mountains is to hide from danger. Actually, the activity also allows people to appreciate the beauty of the autumn. Wearing *zhuyu* plant was also meant to hide from danger. The ancient people wore *zhuyu* plant on their head or placed the plant in a bag.

In 1989, the Chinese government designated the Double Ninth Festival as the Senior's Day. Climbing mountains and showing respect to the elderly have become two customs of the Double Ninth Festival. Drinking Chongyang wine and eating Chongyang cake are two important culinary customs of the festival.

1) Chrysanthemum wine

Chrysanthemum wine is a health preservation drink consumed in autumn. The

---

① zhuyu: A kind of plant.

② danger: Bad luck.

ancient people regarded it as "auspicious wine". Legend has it that there lived a couple in a village. They already had a daughter, but they still wanted a son. Despite years of waiting, their wish hadn't been fulfilled. On the night prior to the Double Ninth Festival, they had the same dream. Someone told them to brew a crock of sweet wine on the ninth day of the ninth lunar month. The wine should be buried underground seven days later and was taken out for consumption before the Spring Festival. If they followed the instructions, the wife would be pregnant three months later. The couple followed the instructions they had received in their dream and they did have a son in the winter of the following year. The child was extremely smart and passed the provincial civil service examination after growing up. Ever since, brewing sweet wine has become a custom of the Double Ninth Festival. On the ninth day of the ninth lunar month every year, the wife of the family would get up early at dawn washing rice and the crock. The husband would chop firewood and carry water. Around 9 a.m., the husband would light up the stove. After boiling the water, the wife would boil the rice and grind the Chinese yeast. After the rice is ready, the wife would pour the cooked rice onto a clean sieve with the rice being spread evenly on the surface of the sieve. The Chinese yeast is then sprinkled on the rice and mixed with the rice. Before the rice cools, the rice is poured into a wine jar and the wine jar is then sealed. Wrap the wine jar with bedsheet or cotton wadding for three to five days. After there is aroma released, the wine jar is buried underground. When the wine jar is taken from underground before the Spring Festival, the wine in the jar is good in color, aroma and taste. The rich aroma released when the jar is opened is really pleasing and invigorating.

2) Chongyang cake

People in some places wear chrysanthemum or eat Chongyang cake on the day of the Double Ninth Festival. Chongyang cake is also known as "flower cake" or "chrysanthemum cake". On the day of the festival, married daughter would return to her parents' home to send Chongyang cake. People also give Chongyang cakes to each other as a gift. Some Chongyang cakes are made with nine layers and in the shape of a pagoda. Two lambs are sometimes added on the top as the Chinese word for "yang" has the same pronunciation with the word for "lamb", hence two lambs symbolize "chong yang" ("chong" means "repeat" or "two"). Some people place a small red paper flag on the cake and light up a candle lamp, probably with the intention of replacing "climbing a mountain" with "lighting up a candle lamp" and "eating cakes".

According to *Notes on Xijing*, a custom of eating lotus cake on the ninth day of the ninth lunar month was established during the Han Dynasty. Lotus cake was the earliest version of Chongyang cake. The custom of eating Chongyang cake became popular during the Song Dynasty. Since the Chinese word for "cake" is pronounced the same with the Chinese word for "high", the custom of eating Chongyang cake symbolizes the custom of "climbing a mountain" during the Double Ninth Festival. In ancient times, the Double Ninth Festival was also an occasion for practicing horse riding, military training and practicing archery. Therefore, some believe that the origin of Chongyang cake was the food rations for soldiers in ancient times.

Chapter 3 Regional Cuisines

China is a country with a long history, vast territory and a large population. China is also a country with rich culinary culture. Over the centuries, due to factors such as geographical conditions, resources, climate and cultural traditions, cuisines of distinctive regional features have been created which are called "regional cuisines". Catering to the taste of people of the region, regional cuisine involves unique cooking methods, seasoning methods, dishes with distinctive flavors, areas covered, history and culture.

3.1 Development and Forming Conditions of Regional Cuisines

China's long culinary history can be traced back to the pre-Qin period. Due to differences in geographical conditions, climate, raw materials, political and economical conditions and customs, different regions developed different cuisines. During the pre-Qin period, there were already southern style cuisine and northern style cuisine. Since the Qin and Han Dynasties, differences between regional cuisines had become more marked. Major traditions of cooking started to be developed in both the southern China and the northern China.[①] During the Tang

① Duli. Comparison of Chinese and Western Cooking History[J]. *Culinary Science Journal of Yangzhou University*, 2002(3): 1-5.

Dynasty, culinary culture saw large-scale development, a result of the rapid development of economy and culture during the period. The creation of high chair and large table made it possible for people to eat at the same table. The southern style cuisine and the northern style cuisine were fully developed during the Tang and Song Dynasties. Unlike the modern times, the northerners preferred sweet food while the southerners preferred salty food during that period. During the Southern Song Dynasty, a large number of Han people moved from the North to the South due to historical reasons and brought with them sweet dishes. As a result, sweet dishes had become the dominant part of the cuisine of the southerners. During the same period, due to the influence from the ethnic minority groups, the northerners living in the Central Plains① started to favor salty dishes, hence today's pattern of "salty dishes in the North and sweet dishes in the South". At the early Qing Dynasty, Sichuan cuisine, Shandong cuisine, Cantonese cuisine and Jiangsu cuisine were the most famous regional cuisines, which were collectively called the "Four Great Traditions". The Four Great Traditions were born in three major cultural areas: Shandong cuisine in the Yellow River cultural area; Sichuan cuisine in the upper reaches of the Yangtze River cultural area; Jiangsu cuisine in the lower reaches of the Yangtze River cultural area; Cantonese cuisine in the Pearl River cultural area.② By the late Qing Dynasty, four new regional cuisine traditions had been developed including Zhejiang cuisine, Fujian cuisine, Hunan cuisine and Anhui cuisine. Along with the Four Great Traditions, the eight cuisine traditions are called the "Eight Great Traditions". After Beijing cuisine and Shanghai cuisine were created, there were "Ten Great Traditions". This chapter focuses on the "Eight Great Traditions" in China and the distinctive Shanghai cuisine.

As traditions of Chinese culinary culture, regional cuisine has to meet requirements on quantity and quality during its development. Forming conditions of

---

① Northern China back then refers to the Central Plains such as areas of Shaanxi and Henan.

② Zhao Jianmin and Jin Hongxia, *Introduction to Culinary Culture of China*[M]. Beijing: China Light Industry Press, 2011: 83.

the "Eight Great Traditions" are as follows:

1. Utilization of local raw materials

Due to the inconvenience of transportation in ancient China, people had little access to raw materials from outside their locality and therefore relied on local raw materials for cooking. As the uniqueness and attractiveness of regional cuisine is largely dependent on the uniqueness of raw materials used and the local distinctive flavor produced, utilization of raw materials is a top priority of preparation of regional cuisine. For example, force-fed duck of Beijing, garlic of Shandong and bean paste of Pixian County in Sichuan are signature raw materials of the respective place, which can't be found in other places.

2. Innovation and originality of cooking techniques

Cooking techniques are one of the major contributors to the creation of a regional cuisine. China has been known for its complicated and skillful cooking techniques. Of the scores of cooking techniques employed in China, each technique is unique. The color, aroma, flavor, appearance and texture of a dish are achieved through the cooking methods used. One small difference may result in different flavor. Some regional cuisines are known for unique flavor because they have some tips for cooking utensils usage, water-based heat utilization, appearance and flavor creation and specific cooking method. Different regional cuisines feature different styles of dishes. For example, Shandong cuisine is known for its soup dishes, Anhui cuisine is famous for its stewed dishes and Sichuan cuisine is famed for its stir-fried dishes. Different cooking techniques will create different flavor and texture of dishes.

3. Numerous famous dishes and snacks promote development of local feast culture

Each of the "Great Eight Traditions" has produced a large number of delicious dishes and snacks. These delicacies and their respective customs have helped promote the creation of local feast culture as well as the creation and development of local feast-based folk activities in the region concerned. As local feast is an opportunity to showcase the local cooking techniques and conduct local folk activities, a feast of varieties of dishes has become an important way of differentiating regional cuisine. Improvement of regional cuisine may also promotes the development of local feast culture.

4. Enduring vitality and potential for development

The creation of regional cuisines in China was not accomplished at one stroke. Rather, it was the result of strenuous exploration over the centuries and contribution of different factors. Only through continuous exploration can the regional cuisines gradually attain perfection. In modern times, different regional cuisines have more opportunities to learn from each other thanks to convenience of transportation and communication. While retaining its own uniqueness, regional cuisine has been improved and enriched through continuous innovation in order to meet diverse demands and tastes, which helps create enduring vitality for the cuisine.

3.2 Main Reasons for Creation of Regional Cuisines

Historical development and distinctive style are two major factors for creation of regional cuisine. Other factors include geographical environment, resources and raw materials, cultural traditions, economic conditions and customs. Among these factors, geographical environment, raw materials and climates play a significant role in the creation and development of regional cuisine.[①]

---

① Zhao Rongguang. *Chinese Culinary Culture*[M]. Beijing: Zhonghua Book Company, 2012: 29 - 30.

1. Climate and geographical conditions

With a vast territory and complex terrain, China is topographically "high in the west and low in the east". The west is made up of high mountains and plateaus. The central part is composed mainly of hills and basins. The east features vast plain. Different geographical location leads to different climate in each region. From the east to the west, the climate changes from moist and semi-humid to semi-arid and arid. Diversity in geographical environment and climate results in regional differences in raw materials used for cooking and culinary preference of local people.

1) Differences in raw materials

As the Chinese saying goes, "those who live on a mountain live off the mountain, those who live near water live off the water", people usually use their local raw materials for cooking. In ancient times, due to inconvenience of transportation and low productivity, people had limited access to raw materials, a fact explaining the formation of fixed culinary cultural areas. For example, people of the coastal area in eastern China eat a lot of seafood because the coastal area is naturally rich in marine products. Shandong cuisine, the foremost of the "Great Eight Traditions", is famed for its seafood dishes. People in the northwest prefer wheaten food because the area is composed mainly of plateau and no sea is in sight, and therefore rice is hardly able to grow in the area. With favorable geographical conditions, the area in the middle and lower reaches of the Yangtze River is dominated by plain and lakes of different sizes. Therefore, rice which grows better in places featuring high temperature and lots of rains is cultivated by the farmers of the area, making the area a fertile "land of fish and rice".

2) Differences in culinary preference

Differences in geographical environment and climate lead to different culinary preference across the country, i.e., "spicy in the east, sour in the west, sweet in the south, salty in the north". The habit of eating peppers is formed due to cold and moist climate. For example, the coastal area in the eastern region is cold and moist in winter. Although Sichuan is not located in the east, the area sees a little sunshine throughout the year since it's situated in a basin. The climate there is therefore cold and moist as well. Eating spicy food may help with perspiration, drive away coldness and dampness and nourish spleen and stomach, and thereby enhancing health of the local people. People in coastal area in the southeast such as Guangzhou and Fujian seldom eat spicy food. They prefer light food since local people are prone to internal heat increase, growing boils, developing phlegm and breathing problems due to hot climate. Light food may help people reduce internal heat.

In the North, it's rather difficult to preserve vegetables in winter due to the coldness. People in the area therefore "pickle" vegetables in order to preserve vegetables for a longer time period. This could explain why most northerners prefer salty food. As for the rainy south, vegetables can grow several rounds throughout the year thanks to its adequate sunshine. With ready access to sweet food items, the southerners understandably have a sweet tooth. Most inhabitants in Guangdong, Zhejiang and Yunnan love sweet food.

Culinary preference of some places has something to do with the water quality of the place. People of Shanxi like vinegar and the province leads other places in the west in terms of "eating sour food". The causes of this practice are because Shanxi is troubled with the problem of insufficient water supply, especially the problem of no access to steady supply of high-quality water. Many places in the province are dominated by low-lying salt marshes and therefore the water and soil have very high alkali content. As a result, the practice of "neutralizing" alkalinity and reducing "water hardness" through eating vinegar has remained popular in the province where a preference for sour food had finally been developed.

2. Historical and cultural factors

Since ancient times, there have been different cultural areas existing in China, namely, the Yellow River cultural area, the Yangtze River cultural area, the Pearl River cultural area and the Liaohe River cultural area. Different areas have developed distinctive cultural features. For examples, the Central Plain is known for its majesty and vastness, gardens in the regions south of the Yangtze River are beautiful and elegant, the southern China with fertile land has flourished for centuries, and the northwest region is famed for its rusticity and simplicity. Different regions have developed different styles of cultural practices. These cultural areas are also the birthplaces of some ancient cities and prosperous business centers in China. For example, Xi'an of Shaanxi, Luoyang and Kaifeng of Henan, Hangzhou of Zhejiang and Nanjing of Jiangsu are famous ancient capitals in China. As the political and economic center of the country in the past, these ancient cities were endowed with a large population and thriving business. Additionally, as the rules of each dynasty tended to pursue culinary enjoyment, court cuisine had been developed which is known for its long-standing history, exquisite preparation process and complicated cooking methods used. Shandong cuisine is the most famous style among all styles of court cuisine. Regions south of the Yangtze River had been known for its beautiful scenery and a large number of men of letters. People there therefore had high expectations for flavor and aroma of dishes and the environment for dining. Over the course of time, Suzhou cuisine featuring dainty food items had been developed. The simplicity of people in the southwest may explain why Sichuan cuisine is simple in style. Sichuan cuisine is a reflection of the charm of the "land of abundance". Having served as a trading port over the years, Guangzhou used many raw materials from outside the city and learned cooking skills from people of other places. This had promoted the refinement of Cantonese cuisine which embodies characteristics of other styles of cuisine.

3. Religious beliefs and customs

In addition to the above-mentioned factors, religious beliefs and customs may also exert influence on the creation of regional cuisines. China is a multiethnic country. The cults and superstitious beliefs of a certain ethnic group may somewhat determine what raw materials are chosen for cooking and how dishes are cooked. Religious beliefs of local people also determine what materials are used for cooking. Since ancient times, culinary traditions had been intertwined with religious practice. For example, after Buddhism was introduced into China, the practice of vegetarian diet was then spread across the country and in turn drove the cultivation of vegetables, fruits and development of skills for making bean products. As a result, a vegetarian cooking style was created and became one part of the culinary culture of China. Ethnic customs and individual preferences also play an important role in development of culinary traditions. For example, sachima of Manchu, kabob of Uygur and bamboo tube rice of the Dai people represent the culinary traditions of different ethnic group. Since people of Huizhou drink tea all year round, their dishes are usually greasy, i.e., their dishes are "greasy, deep in color and with strong flavor". People in the western region prefer hard to chew food, which easily causes indigestion. To help with digestion, people there eat a lot of sour dishes, an example of how individual preference has exerted influence on creation of regional cuisine.

4. Physiologic and psychological rejection

The Chinese nation places great value on history, family and tradition. While preserving traditions inherited from their ancestors, every region or ethnic group has developed their own unique customs. They hold fast to their culinary traditions, seldom wanting to change them. Such a psychological factor leads to deeply-ingrained culinary traditions of individual region. Consumption of a certain type of food for a long time may cause changes in digestive organ and consequent physiologic rejection. When a northerner goes to the South and eat rice, he or she may still feel hungry after eating, unlike steamed bun, rice can't expand in the stomach. Those who normally eat vegetables may suffer from indigestion if eating meat for several days in a row. Due to these factors, each regional cuisine has been able to retain its unique style.

Where Does the Term "Drinking Vinegar" Come From?

"Vinegar" is one of the basic seasoning for Chinese cooking. Due to water quality problem, people of Shanxi like drinking vinegar. Connotation was then created from the term "drinking vinegar" meaning "jealousy", i.e., if a third party is involved in a romantic relationship between a man and a woman, those involved may start "drinking vinegar" or burning with jealousy.

Then, what is the connotation of the term "drinking vinegar"? Here is the story: Li Shimin, Emperor Taizong of Tang Dynasty decided to find concubines for Fang Xuanling, the Prime Minister so as to win more support from Fang Xuanling. However, the wife of Fang Xuanling didn't want him to accept the arrangement because she burned with jealousy. Finally, Emperor Taizong had no other choice but to order Mrs. Fang to choose between two options: drinking poisonous wine or allowing her husband to have concubines. However, the strong-willed Mrs. Fang would rather die than yield to the pressure from the emperor. She then took the cup and gulped down the "poisonous wine". After Mrs. Fang drank the wine with tears in her eyes, she found out that what she had drunk was actually strong and flavorful vinegar rather than poisonous wine. Ever since, "becoming jealousy" and "drinking vinegar" share the same meaning with the latter becoming the metaphor for the former, a usage remaining popular today.

3.3 Regional Cuisines

1. Shandong cuisine

Lu cuisine, or Shandong cuisine, is the foremost among the culinary traditions of Chinese cuisine. The development of Lu cuisine is closely linked to the geographical conditions, history, culture, economic conditions and folk customs of Shandong. Geographically speaking, Shandong is one of the birthplaces of Chinese ancient civilization. Located in the lower reaches of the Yellow River, Shandong enjoys a mild climate. The Jiaodong Peninsula lies between the Bohai Sea and the Yellow Sea. Thanks to the mountains, rivers, lakes and vast fertile land in Shandong, the province is rich in produce ensuring a steady supply of raw materials for culinary endeavors. This has served as an important foundation for establishing Shandong cuisine's status as the No. 1 regional cuisine in the country. Culturally speaking, as the "hometown of Confucius and Mencius", Shandong enjoys rich cultural heritage. The Confucian culture has facilitated the publicity efforts of Shandong cuisine whose popularity is incomparable. Politically and geographically speaking, Shandong was closest to the power center of the Central Plains which constituted a geographical advantage for Shandong cuisine to have become the palace cuisine in feudal China.

The history of Shandong cuisine can be traced back to the Spring and Autumn period and the Warring States period. During the Western Zhou Dynasty, Qin Dynasty and Han Dynasty, Qufu, capital of the State of Lu, and Linzi, capital of the State of Qi, were both prosperous cities. As a result, catering industry flourished and famous chefs emerged in large numbers. Since the Song Dynasty, Shandong cuisine has become the representative of the "north cuisine". During the Ming and Qing Dynasties, Shandong cuisine became palace cuisine and the foremost of the "Eight Great Traditions". It's also one of the cuisines which involve numerous cooking methods, complicated preparation process and advanced skills.

Shandong cuisine is known for its aromatic, tasty and authentically flavored dishes. Much effort is made to make clear soup and creamy soup. Clear soup is fresh and creamy soup is rich in flavor. Shandong cuisine is made up of three styles: Shandong style, Jiaodong style and Confucius Mansion style. Shandong style dishes are aromatic, tasty and authentically flavored. Soup, especially stock making is one distinctive feature of Shandong style. Fushan of Yantai is the birthplace of the

Jiaodong style which is known for its seafood dishes. Jiaodong style usually offers light dishes which are rich in contents. With Qufu dishes as the representative, Confucius Mansion style of dishes are popular in southwest Shandong and Henan. It's famed for its exquisite preparation process, extensive use of raw materials and varieties of dishes offered. Shandong cuisine typically uses 30 cooking methods with quick-frying and braising as the most commonly used methods. Quick-frying uses big fire for quick cooking while braising is a method unique to Shandong cuisine.

Representative dishes of Shandong cuisine are: Dezhou braised chicken; fried carp with sweet and sour sauce; braised sea cucumber with scallions; braised intestines in brown sauce.

Dezhou braised chicken

Description: five-fragrant and boneless; tender, tasty and savory; chicken's drumsticks are bent upwards towards the head and chicken feet are inserted into chicken breast; chicken wings pass through chicken neck and then come out of the mouth in a crossed manner; chicken is served in supine position covered with golden luster.

Dezhou braised chicken

Fried carp with sweet and sour sauce

Description: Fish tail raised; amber colored; crispy outside and tender inside; in the shape of carp jumping over the dragon gate; sour and sweet; symbolizing the happy events such as passing civil service

Fried carp with sweet and sour sauce

examination and getting married.

Braised sea cucumber with scallions

Description: fresh, tender and tasty; scallions with rich aroma; a cholesterol-free dish rich in collagen protein; healthy and nutritious food.

Braised sea cucumber with scallions

Braised intestines in brown sauce

Description: ruddy color; tender texture; sour, sweet, bitter, spicy and salty; a rare five-taste dish in Chinese cuisine.

Braised intestines in brown sauce

2. Anhui cuisine

Hui cuisine or " Anhui cuisine" is one style of Chinese cuisine whose development had been based on the success of Huizhou businessmen. The rise and fall of " Anhui cuisine" had depended on the ups and downs of the career path of Huizhou businessmen. The rise of Huizhou businessmen drove the development of the catering industry of Anhui as they tended to demand dainty food after their success. As a result, Anhui cuisine had undergone improvement and became an indispensable part of a banquet. Anhui restaurants are opened across the country increasing the popularity of Anhui cuisine. The creation of Anhui cuisine depended largely on the unique geographical environment, cultural heritage and culinary traditions of Huizhou in the old times. Huizhou is known for its beautiful scenery, numerous gullies and pleasant climate. Thanks to these favorable conditions and the steady supply of raw materials, Anhui cuisine has developed continuously over the years. At the same time, various kinds of customs and seasonal activities have also promoted the creation and development of Anhui cuisine.

Features of Anhui cuisine include: obtaining raw materials locally; delicious and savory dishes; relying on water-based heat; greasy dishes with deep color; rich and mellow taste; authentic flavor. Huizhou is rich in fresh fish and shrimps and poultry. The practice of obtaining raw materials locally ensures the distinctiveness of Huizhou cuisine as well as the freshness of dishes. Using natural materials and emphasizing dietetic invigoration are two prominent features of Anhui cuisine.

Representative dishes of Anhui cuisine are: stinky mandarin fish; steamed partridge; Wenzheng mountain bamboo shoots.

Stinky mandarin fish

Description: stinky yet flavorful; tender and savory; retaining the authentic flavor of mandarin fish.

Stinky mandarin fish

Steamed partridge

Description: clear soup, rich aroma; tender, tasty and authentic in flavor.

Wenzheng mountain bamboo shoots

Description: fragrance of cured meat and bamboo shoots; hot dish; oily but not greasy; salty and sweet; flavorful and tasty.

Wenzheng mountain bamboo shoots

3. Zhejiang cuisine

Zhejiang cuisine is made up of a number of styles of cooking such as Hangzhou style, Ningbo style and Wenzhou style. Unlike Shandong cuisine which has been influenced by historical and political factors, Zhejiang cuisine is known for its cultural heritage. As the ancient saying goes, "Just as there is paradise in heaven, there are Suzhou and Hangzhou on earth." The birthplace of Zhejiang cuisine is located in this area which is endowed with beautiful scenery and rich resources, especially fish and shrimps. Dishes of Zhejiang cuisine can be described as light, aromatic, crispy, tender, refreshing and fresh. Raw materials used are fresh and tender. Ingredients usually include fish, shrimps, poultry, farm livestock and seasonal vegetables. The skill of cutting up raw materials is emphasized. Dishes are light, fresh and authentic in taste. Many famous dishes were created by the common people and dishes are small in size. Dishes are named after the surroundings. Ningbo style is known for its seafood dishes which are fresh and tender in taste. Cooking methods include frying, steaming, stewing and pickling. Soup and stock making is emphasized as well as the authentic flavor of raw materials. Similar to Ningbo style, Wenzhou style offers many seafood dishes. Cooking methods feature "two less, one emphasizing", i.e., less oil, less thicken soup and emphasizing cutting skills. As a result, Wenzhou style has created many dishes with unique flavor.

Representative dishes of Zhejiang cuisine are: West Lake sweet-and-sour fish; Dongpo pork; stir-fried beancurd rolls stuffed with minced tenderloin; West Lake water shield soup; fried shrimps with Longjing tea; Ningbo sweet dumplings.

West Lake sweet-and-sour fish

Description: strict selection of materials, normally 1.5 jin of grass carp; strict rules on duration and degree of heating during cooking, normally cooking three to four minutes; sprinkle with sugar and vinegar when ready; dish covered with red cluster; tender with a slight taste of crab; sour and sweet; tasty and unique in style.

West Lake sweet-and-sour fish

Dongpo pork

Description: thin skin and tender meat with red cluster; rich in flavor and thick gravy; crisp yet shape retained; sticky, aromatic but not greasy; excellent in color, aroma and taste; a popular dish.

Dongpo pork

Stir-fried beancurd rolls stuffed with minced tenderloin

Description: thin skin of soya-bean milk stuffed with minced tenderloin and then fried; dish covered with yellow luster; with the appearance of horse bell; crispy and refreshing.

Stir-fried beancurd rolls stuffed with minced tenderloin

Tale about West Lake Sweet-and-Sour Fish

The dish of "West Lake sweet-and-sour fish" is also known as "story of sister-in-law and brother-in-law". Legend has it that there were two brothers with the surname of Song used to live in Hangzhou during the Southern Song Dynasty. Both of them were learned men. But they didn't want to secure an official position and serve for the government. They then lived in seclusion in a mountain forest and made a living by fishing. The elder brother's wife was very beautiful. Coveting her beauty, a local tyrant killed the elder brother so as to forcibly marry Mrs. Song. Facing the unexpected adversity, Mrs. Song and the younger brother were extremely grieved. They filed a lawsuit against the local tyrant but failed to bring him to justice. Instead, they were cruelly beaten. After returning home,

Mrs. Song urged the younger brother to flee to other place and make a living there. Before the younger brother left, Mrs. Song used sugar and vinegar to make a dish of fish. She then said to the younger brother, "This dish is both sour and sweet. May you someday become an outstanding man. But never forget today's adversity." The younger brother then left home and joined the army to resist Jin's invasion. He finally became a high-ranking official. After returning to Hangzhou, he punished the local tyrant severely. However, he knew nothing about the whereabouts of his sister-in-law. Conscience-stricken, he made every effort to try to find his sister-in-law. One day, when he had meal at a restaurant, he found that the taste of a dish of fish was the same as the one his sister-in-law had cooked for him. After inquiry, he finally found his sister-in-law who had been working at the restaurant by concealing her identity. The story ends with their happy reunion.

4. Jiangsu cuisine

Su cuisine or Jiangsu cuisine is the representative style of cooking in the middle and lower reaches of the Yangtze River. With a long history and rich heritage, Jiangsu cuisine features culinary preference of people of the regions south of the Yangtze River. Jiangsu cuisine is made of different styles of cooking including Suzhou style, Yangzhou style, Huai'an style and Nanjing style. Huaiyang cuisine is the most prestigious style of Jiangsu cuisine. Huaiyang cuisine was originated from the ancient city of Yangzhou and Huai'an. The two places are known for their prosperity, large numbers of men of letters and abundance of resources, especially aquatic products. Therefore, Jiangsu cuisine is also known as Huaiyang cuisine.

Features of Jiangsu cuisine are: strong flavor yet slightly light; aromatic, tasty and tender; authentically flavored; oily yet not greasy; mild taste; salty and sweet. Cooking methods include stewing, simmering, braising and frying. Raw materials are carefully selected and special attention is paid to color and appearance. Different season offers different kinds of dishes. Suzhou style is known for its sweet-flavored dishes and attractive color combination. Yangzhou style offers light and refreshing dishes with plenty of main ingredients. Yangzhou style emphasizes cutting skills and rich flavor of dishes. Dishes of Nanjing style are mellow in taste and small in size. Duck-based courses are the most popular dishes in Nanjing style of cooking.

Representative dishes of Jiangsu cuisine are: squirrel-shaped Mandarin fish; braised ham in clear soup; Yangzhou fried rice; pork balls with crab sauce; boiled salted duck.

Squirrel-shaped Mandarin fish

Description: squirrel-shaped; crispy outside and tender inside; orange in color; sour-and-sweet and tasty.

Squirrel-shaped Mandarin fish

Braised ham in clear soup

Description: clear soup; rich flavor; tasty ham; employing advanced "stock making" techniques invented by chefs of Yangzhou.

Braised ham in clear soup

Yangzhou fried rice

Description: grains are distinctive and loose; appropriate degree of softness and hardness; pleasing color with luster; various ingredients; tasty and refreshing.

Yangzhou fried rice

5. Sichuan cuisine

Chuan cuisine or Sichuan cuisine constitutes an important part of Bashu culture, which originated in the State of Ba and the State of Shu. As the most unique cuisine, Sichuan cuisine is rather popular nowadays. With a long history, Sichuan cuisine is well-known at home and abroad. Today, Sichuan restaurants can be found in every corner of every city in China, hence the name "food for common people". Development of Sichuan cuisine is closely linked with the unique natural and geographical conditions of Sichuan. Located in the middle and lower reaches of the Yangtze River, Sichuan is surrounded with hills resulting in a warm and moist climate. Abundance of raw materials is available in Sichuan. Due to the moist climate, people of Sichuan usually eat peppers to remove dampness from the body and thereby preserving health.

Chengdu style and Chongqing style are the two representative styles of Sichuan cuisine. Sichuan cuisine features strict selection of raw materials, large size of dishes, distinctiveness of ingredients and ingredients collocation. Sichuan cuisine relies on a wide range of raw materials and heavily uses hot seasoning. The three "peppers" (chilli, black pepper and Chinese red pepper) and fresh ginger constitute an indispensable part of Sichuan cuisine. Typical dishes of Sichuan cuisine are spicy, sour and numbing. Sichuan-style dishes are renowned for their various tastes and each dish has a particular taste. A total of 23 tastes have been created including home-made taste, salty and delicious taste, fish-flavored taste, lychee taste and odd taste. Commonly used cooking methods include baking, braising, dry stir-frying and steaming.

Sichuan cuisine has earned international acclaim ("delicacies in China, tasty food in Sichuan"). Representative dishes of Sichuan cuisine are: Yuxiang shredded pork sauteed in spicy garlic sauce; spicy diced chicken with peanuts; Mapo tofu; sliced beef and ox organs in chili sauce; twice-cooked pork slices; DongPo stewed pork joint.

Yuxiang shredded pork sauteed in spicy garlic sauce

Description: strict selection of raw materials; ruddy color; fish flavor; salty, sweet, sour and spicy in taste; tender shredded pork.

Yuxiang shredded pork sauteed in spicy garlic sauce

Spicy diced chicken with peanuts

Description: spicy and sweet; tender chicken and crispy peanuts; spicy and tasty; red in color; mildly spicy; tender chicken.

Spicy diced chicken with peanuts

Mapo tofu

Description: smooth texture; white and tender tofu; numbing and flavorful in taste.

Mapo tofu

Origin of Mapo Tofu

During the reign of Tongzhi in Qing Dynasty, there was an eatery near Wanfuqiao Port in Chengdu. The wife of the owner of the eatery had pock marks on her face and therefore was called "Chen Mapo". The eatery regulars included dockers and porters. One day, when the eatery was about to close, a group of men came inside asking the owner to make them some dishes which would be hot, cheap and go well with rice. Looking around, Chen Mapo only found several plates of tofu and a small amount of minced beef. And it was too late to go to the market to buy raw materials. What was she supposed to do? Chen Mapo suddenly had an idea. She chopped beans in tiny bits and fried them in oil along with fermented soya beans. Adding more water, she poured tofu into the wok. The tofu had been cut into pieces with the length of a finger. She then added other seasoning, fried minced beef and chicken soup to make a dish, which was then topped with Chinese red pepper powder. A dish of tofu with numbing, spicy, hot, tender and aromatic taste was then served. The men sweated while eating their fill. At the same time, they all said it's a really tasty food. The news spread and finally many people knew that tofu dish made by Chen Mapo was delicious and cheap and went well with rice. More people came to the eatery to order the dish which then became a signature dish of the eatery. Since the dish was numbing and spicy and the name of the owner's wife was Chen Mapo, this dish of tofu finally got the name of Chen Mapo tofu.

6. Hunan cuisine

Xiang cuisine or Hunan cuisine enjoys a long history which was created during the Han Dynasty. Located in the central south of China and the middle reaches of the Yangtze River, Hunan has a warm climate and abundant rainfall. Similar to Sichuan, Hunan has a warm and moist climate due to its geographical position. Therefore, people in Hunan eat a lot of peppers in order to refresh themselves and remove dampness.

Hunan cuisine is based on cuisines of areas along Xiangjiang River and the Dongting Lake as well as cuisine of western Hunan. Hunan cuisine emphasizes interactions among different raw materials whose aroma may be mixed together. Hunan cuisine usually offers sour and spicy dishes. Representative dishes of Hunan cuisine are either spicy or smoky. Cooking methods often used include curing, smouldering, simmering, steaming, stewing and frying. Representative dishes are: steamed fish head with diced hot red peppers; smoky flavors steamed together; Dong'an chicken; lotus seeds in rock sugar syrup.

Steamed fish head with diced hot red peppers

Description: Palatable fish head and spicy chopped peppers cooked together; white and fresh fish head is covered with red chopped peppers releasing aromatic steam; aroma of fish head is retained by steaming and the spicy flavor of chopped peppers penetrates into fish head; tender and tasty.

Steamed fish head with diced hot red peppers

Smoky flavors steamed together

Description: rich smoky taste; sweet and salty tastes well-balanced; red luster; tender and not greasy; a small amount of thick gravy; flavors of ingredients combined contribute to the taste of dish.

Smoky flavors steamed together

Dong'an chicken

Description: attractive appearance; bright color; tender meat; sour and spicy; refreshing; oily yet not greasy; nutritious.

7. Cantonese cuisine

Guangdong cuisine or Cantonese cuisine has developed at a fast pace despite its somewhat late creation. Cantonese restaurants can be found in many places in China. Chinese restaurants in a lot of foreign countries mostly offer Cantonese dishes. The geographical environment of Guangdong has played a significant role in the development of Cantonese cuisine. Located in the coastal area in southeast China, Guangdong is known for its hot and moist climate. With complex terrain and numerous islands along the coast, Guangdong is rich in rare seafood, usually in a wide variety.

Cantonese cuisine features extensive use of various kinds of raw materials. Due to climate concerns, Cantonese cuisine usually offer seasonal dishes with light dishes in summer and autumn and rich flavor dishes in spring and winter. Advocating dietetic invigoration, dishes of Cantonese cuisine usually offer heavy, light and refreshing taste. Cooking methods often used include frying, deep frying, stewing and braising. With rich color, dishes are flavorful yet not greasy. Cantonese cuisine also offers delicate dim sum and sumptuous dishes. Representative dishes are: roasted suckling pig in open oven; Guangzhou Wenchang chicken (sliced chicken with chicken livers and ham); plain boiled chicken; Baiyun trotter; fried beef in oyster sauce.

Roasted suckling pig in open oven

Description: red in color with shimmering luster; crispy outside and tender inside; unique taste and style.

Guangzhou Wenchang chicken

Description: attractive appearance; smooth texture; thin skin and tender meat; aromatic; oily yet not greasy.

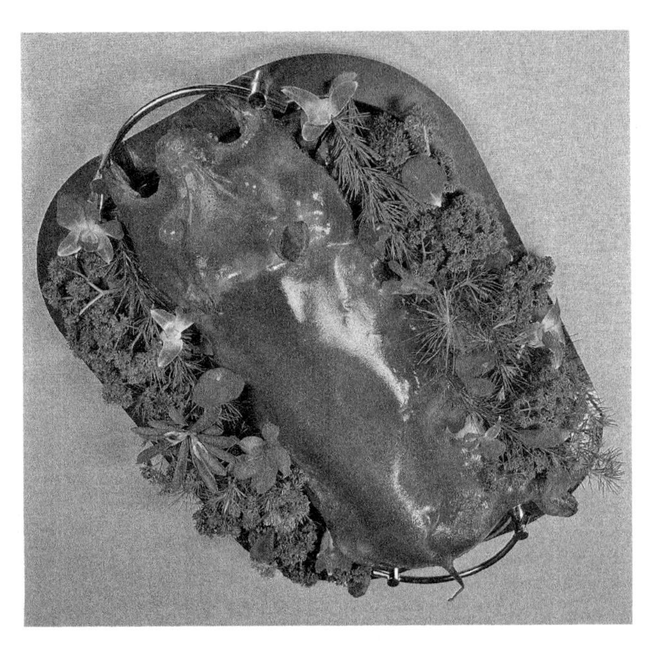

Roasted suckling pig in open oven

Guangzhou Wenchang chicken

Plain boiled chicken

Description: white in color with a yellowish tinge; aroma of fried scallions; scallions pieces with flower edge; preferably accompanied by mustard and soy sauce.

Plain boiled chicken

8. Fujian cuisine

Min cuisine is the abbreviation of Fujian cuisine. Fujian is a well-known hometown of overseas Chinese who have continuously introduced new kinds of food and seasoning into Fujian. These food and seasoning are integrated with the traditional cooking styles of China enriching the culinary culture of Fujian and creating a unique cooking style which is ready to embrace foreign influences. Fujian cuisine has developed on the basis of cuisines of Fuzhou, Quanzhou and Xiamen. Features of Fujian cuisine include: use red fermented rice paste as seasoning; superb soup making skills; emphasizing cutting skills. As for the cutting skills of chefs of Fujian, there is a description which says, "sliced food is as thin as a piece of paper, shreds are as thin as hair, carved food is as beautiful as lichee." Dishes of Fujian cuisine are beautiful in color and palatable in taste. Cooking methods commonly used include stir-frying, stir-frying with thicken soup, frying and simmering. Using red fermented rice paste as seasoning is a unique method employed in Fujian cuisine. Representative dishes of dishes using red fermented rice paste include fish in wine sauce, pork in wine sauce and chicken in wine sauce.

Fujian cuisine inherited fine cooking techniques from Chinese cuisine and at the same time, it has developed its own style typical of the southern China. Dishes of Fujian cuisine are fresh, tasty, rich, aromatic, oily, sour-and-sweet in taste. Representative dishes are: Buddha jumps over the wall; deep-fried shrimps with shredded ham; chicken in wine sauce; fried prawn shaped as a pair of fish; lychee meat.

Buddha jumps over the wall

Description: Many ingredients are simmered together sharing meat taste of each other and retaining taste of individual ingredient; tender and smooth texture; rich in meat taste; oily yet not greasy; flavors of different ingredients mixed enhancing the overall taste; nutritious and capable of nourishing qi and blood, clearing lung and moistening intestines and preventing asthenia cold.

Buddha jumps over the wall

Lychee meat

Description: with a red tinge; shaped like lychee; crispy and tender; aromatic; sour and sweet.

Lychee meat

Chicken in wine sauce

Description: red tint; crispy bones; tender meat; aromatic; oily yet not greasy.

9. Shanghai cuisine

Shanghai cuisine is the local cuisine of Shanghai. A prominent feature of Shanghai cuisine is heavy use of oil and soy sauce (oily, rich flavor, sweet, bright color). Commonly used cooking methods include braising in soy sauce and simmering. Sugar is heavily used for cooking. Dishes are salty and sweet, oily yet not greasy.

Shanghai cuisine was initially targeted at common people and therefore didn't appeal to refined taste. It then learned from other regional cuisines especially Jiangsu cuisine. By the middle of the 20th century, Shanghai cuisine had developed its distinctive features such as selection of fresh materials and mild taste. Many Shanghai restaurants created their signature dishes and produced a large number of celebrated chefs, and thereby elevating the status of Shanghai cuisine. Now, to follow the worldwide culinary trend of consuming less sugar, fat and sodium, dishes of Shanghai cuisine also contain less oil and sugar so as to adapt to taste of people of today. In addition to dishes cooked with large amount of sugar and soy sauce, Shanghai cuisine also offers seasonal dishes which are light in taste. Using red fermented rice paste as seasoning is a method frequently used in Shanghai cuisine. Dishes such as chicken in wine sauce, pig's trotter in wine sauce and green soy beans in wine sauce are unique to Shanghai cuisine.

Representative meat dishes of Shanghai cuisine are: oil ring eel paste; braised black carp tail in brown sauce; fried shrimps; braised river crab with soy paste; braised pork intestines; pig tripe and chayote; braised longsnout catfish in brown sauce; braised chestnut chicken with yellow sauce. All these meat dishes exhibit the prominent feature of Shanghai cuisine, i.e., "heavy use of oil and soy sauce". Vegetable dishes are usually seasonal dishes such as Indian kalimeris herb, shepherd's purse, Chinese little greens and Shanghai edible rape. All these vegetable dishes are refreshing in taste. Snacks of Shanghai cuisine include small steamed bun stuffed with pork and moon cake stuffed with pork.

Braised pork intestines

Description: covered with golden luster; thick gravy; sticky, aromatic and oily; tender as gluten; salty and slightly sweet.

Braised pork intestines

Oil ring eel paste

Description: palatable eel meat; rich aroma; tasty and capable of stimulating appetite; nutritious.

Oil ring eel paste

Braised black carp tail in brown sauce

Description: black carp tail braised in brown sauce; red luster; thick gravy; oily and tender.

Braised black carp tail in brown sauce

Chapter 4 Snacks

4.1 Cultural Background of Snacks

China is a country with a vast territory and numerous ethnic groups. Snacks of distinctive local features and ethnic flavors can be found in every corner of the country. As an important constituent of Chinese culinary culture, snacks enjoy a special status on Chinese peoples' dining table. Most snacks were first created by street pedlars by using local specialties as raw materials. Snacks were usually sold on the street readily available for purchase. Snacks are known for distinctive local style and flavor.

Similar to culinary culture, evolution of snacks culture is also the result of differences in geographical conditions and historical progress.

1. Natural environment

Natural geographical environment determines types of crops cultivated and in turn affects culinary features. Unique natural geographical environment can protect the development of culinary culture. A typical example is the differences in culinary culture between the South and the North. Differences in geographical environment between the South and the North lead to different types of crops grown in these two regions. Over time, due to these differences, different culinary traditions and preferences have been developed in different regions, a cause of the markedly different culinary traditions in the South and the North. The culinary culture of the wheat growing regions with the Yellow River basin as the representative and that of the rice growing region with the Yangtze River basin and the area to its south as the representative shows significant differences. As a result, snacks of the northern regions are mostly made of wheat while most snacks of the southern regions, especially those of the Yangtze River basin, are made of rice. For example, pita bread soaked in lamb soup, buckwheat noodles and Qishan minced noodles are all made of wheat. On the other hand, Ningbo sweet dumplings, Shanghai pork ribs with rice cakes and Jiaxing zongzi are made of rice.

2. Climatic conditions

Climate and soil property play a role in the development of culinary preferences. For example, due to the moist and cold climate, people in Sichuan, Guizhou and Hunan eat a lot of spicy food in order to dispel cold and eliminate rheumatism. Due to the high content of calcium in the soil of the Loess Plateau in Shanxi, people there favor sour food with a view to prevent calcium deposition in the body and calculosis. Many ethnic groups in Guizhou prefer sour food; this is in part because of the annual rainy season in the middle and lower reaches of the Yangtze River as well as the incessant rainfall caused by the mountain climate of Guizhou. Both the Dandan noodles (Sichuan noodles with peppery sauce) and tasty lobster of Hunan are numbing and spicy in taste. Sour soup powder and pickled vegetables in cold sauce are two examples of Guizhou people's preference for sour food. Additionally, people in Shanxi eat vinegar for every meal. All these reflect how climate and soil affect people's culinary preference and how people adapt to natural environment through dietetic invigoration.

3. Cultural environment

Creation of local snacks have also been influenced by cultural environment such as political and economic environment and religious/ethnic environment.

1) Political and economic environment

Snacks had appeared during the pre-Qin period. With the growth of commercial economy, snacks had undergone rapid development and gained wide popularity. They first appeared in major political, economic and cultural centers such as the prosperous cities. These ancient metropolises normally enjoyed political stability, large population concentration and booming economy. Additionally, pursuit of culinary pleasure by the rulers and the merchants, development of imperial cuisine, banquets thrown by the government officials and parties of the scholars all greatly promoted the improvement of local cooking techniques.

2) Religious and ethnic environment

Among all the religions practiced in today's China, only Taoism was born indigenously. All other religions including Buddhism, Islamism, Catholicism and Christianity were introduced from outside. Religiously speaking, Buddhism and Islamism had been the two greatest factors in the development of culinary culture. China has a total of 56 ethnic groups. Different culinary traditions of individual ethnic group have led to the creation of different styles of local snacks. However, due to such reasons as primitive worship and taboo, people living in the same region may create different types of snacks. For examples, both the Manchu people and ethnic Koreans are native to Northeast China. However, although they live in same natural and climatic environment, they have different culinary traditions. The Manchu people never eat dog meat because they love dogs. Most Manchu people rely on hunting for a living and dogs play an indispensable role in their hunting activities. Naturally, dog meat had gradually disappeared from their dining table. Tales about how a loyal dog saves its owner are also often told by the Manchu people. Under Nurhachi, Emperor Taizu of the Qing Dynasty, prohibition against eating dog meat became an unwritten law. On the other hand, dog meat is an important food item on the dining table of the ethnic Korean who lives in the same region. Therefore, food taboos originating from religious rules have led to different culinary culture between different ethnic groups who inhabit the same regions. At the same time, the same ethnic group has developed the same culinary culture even their members live in different regions. For example, the Manchu people and the ethnic Koreans living in Northeast China have developed different culinary culture. Some ethnic groups live in the same regions of Qinghai and Gansu also has different culinary preferences while the Hui people found throughout China share the same culinary culture.

3) Cultural integration

During the course of the development of culinary culture in China, ethnic groups have started to adopt culinary traditions of other ethnic groups. For example, the Kazakhs living in Xinjiang mainly have tea, meat (Naren[①] and smoked horse sausage), milk and wheaten food for their daily diet. Except following these culinary habits, those Kazakhs who live in the cities have also learned to make different delicacies from other ethnic groups and thereby consuming a wider variety of food. They in turn influenced the Kazakhs living in farming and pasturing areas who then had access to more food choices. For example, rice sausage of the Uygurs, mutton/horsemeat/beef pilaf of the Uzbeks and vermicelli soup of the Hui people have all became one part of the daily diet and snacks of the Kazakhs.

4.2 Development of Snacks

Development of snacks in China was the result of an increasingly flourishing

① Naren is a delicacy of pasturing area of Xinjiang. With a distinctive style of pasturing area, the delicacy is also called hand-grabbed meat and hand-grabbed mutton noodles.

catering industry. The term "snacks" appeared in *Rites of Zhou* and *Evocation*, two books devoted to the pre-Qin period. During this period, snacks were normally not counted as one component of a meal although they can be put on dining table as "dim sum". Snacks can also be eaten on the go. However, snacks during this period were not accessible to the general public. Catering industry thrived during the Han Dynasty ("Delicacies can be found in every corner of a city"). Snacks sold in shops included sweet soybean milk, braised meat seasoned with soy sauce and cooked chicken. In "Biography of Important Merchants" of *Records of the Grand Historian*, there is an anecdote about how a man makes a fortune by selling snacks such as soybean milk and becomes as rich as a duke.

During the late Han Dynasty, fermentation technology was employed for production of wheaten food which then gained wide popularity. Wheaten food such as steamed cake, Hu cake (baked wheat flour cake topped with sesame), steamed bun and jiaozi had been created. Both as items of formal meal and snacks, these wheaten food items were cheap, easy to carry around and available for purchase in shops. All these contributed to their popularity. During the Tang and Song Dynasties, more varieties of snacks were created and consumed by people of different social classes. "Flower cake" was introduced from the imperial palace to the common people and was also sold in shops. There were also seasonal snacks which can meet needs of the common people. Snacks shops were found everywhere such as Hu cake shop, steamed cake shop, wonton shop and steamed stuffed bun shop. There were also food vendor's stands such as steamed cake stand run by Wu Dalang as described in *The Water Margin*, a popular fiction. Morning market snacks, midnight snacks and snacks for children were new forms of snacks having appeared over time. Snacks brands were also created such as Cao Fafa meatpie and Taisui steamed bun. During the Ming and Qing Dynasties, snacks had undergone further development. Today, economic growth brings higher living standards to people and the importance of dining has decreased. Besides, busy work schedule also makes it impossible for people to have formal meals. Consequently, they have snacks instead. Some cities have established fold culture centers where various kinds of snacks can be found. For example, Xisi, Dashanlan, Tianqiao and Wangfujing of Beijing, Nanshi food street of Tianjin, Shanghai Chenghuang Temple, Suzhou Xuanmiao Temple, Chongan Temple of Wuxi and Nanjing Confucius Temple are famous snacks markets which have been in existence for several hundred years. These snacks markets exhibit distinctive features of culinary culture of common people and thereby constituting important tourism resources which have their unique charm.

4.3 Classification of Snacks

1. Geographical classification

As the third largest country in the world by land area, China has a vast territory, which leads to huge differences in culinary culture between different regions and consequently a rich culinary culture of the country. Each region in the country has created its own unique dishes and at the same time, snacks with distinctive local flavor. These snacks reflect not only the local culinary culture but also local customs and local folk culture. Geographically, local snacks in China can be classified into eight major groups.

1) Style of Beijing-Tianjin

Due to the unique historical status of Beijing and Tianjin and the fact that they are home to culinary cultures of different ethnic groups, Beijing-Tianjin snacks hold a special place among local snacks in China.

A. Beijing snacks

It's estimated that Beijing has created 3,000 kinds of snacks over the centuries. Several factors have contributed to this amazing achievement such as unique historical status of Beijing, long history of the city, creation of different ethnic groups, traditional festivals and religious practices, access to a wide choice of raw materials and various cooking techniques and methods used. Today, Wangfujing Snacks Street, Qianmen Street and South Luogu Lane are must-go destinations for visitors to Beijing. Most famous snacks of Beijing are: rolling donkey (soybean-flour cake), fermented bean drink and pea flour cake.

Rolling donkey (soybean-flour cake)

Rolling donkey is one of the traditional snacks in the North. With yellow, white and red in color, the cake is attractive in appearance. When the cake is topped with soybean flour as the final step of production, it looks like the dust rose when wild ass rolls on the ground in the suburbs of old Beijing, hence the name "rolling donkey".

Rolling donkey

Ingredients of "rolling donkey" include rhubarb rice flour, soybean flour, sweetened bean paste, white sugar, sesame oil, osmanthus, green & red condiments and melon kernels. The process of making rolling donkey consists of three steps: making cake base, mixing ingredients and refining the shape. Covered with soybean flour, finished "rolling donkey" cake is covered with golden yellow luster. The fillings are sweet with soybean aroma. The soft and sweet fillings melt with just one bite. There is no need to chew the soybean flour when eating. As a traditional snack suitable for all ages, "rolling donkey" is a delicacy you don't want to miss.

Fermented bean drink

Fermented bean milk is a traditional snack unique to the old Beijing. It was first created as a beverage for the imperial household of the Qing Dynasty. According to written records, the snack has a 300-year history.

Fermented bean milk uses mung beans as raw materials. After starch is filtered to make vermicelli and other food items, the residue would be fermented to make fermented bean milk. Although with a peculiar smell, fermented bean milk possesses many health preservation properties such as nourishing the stomach, detoxification and reliving internal heat. When drinking in summer, it can quench thirst and relieve summer heat. When drinking in winter, it can clear heat and warm yang. If drinking it throughout the year, it can stimulate the appetite, strength the spleen, detoxicate the body and remove dryness. Inhabitants of old Beijing usually accompanied fermented bean milk with fried ring and salted vegetables shreds. Fermented bean milk is one component of the unique culinary culture of the old Beijing.

Pea flour cake

Pea flour cake is a seasonal snack consumed in spring and summer in Beijing. Originally a snack for common people, it was then introduced to the imperial household. Pea flour cake made by the imperial household during the Qing Dynasty selected first-rate white peas as raw materials. The finished pea flour cake is pale yellow in color with fine and pure texture. It melts in mouth and sweet and refreshing in taste. It became famous because Empress Dowager Cixi took a fancy to it. Steps for making pea flour cake include: grounding peas, peeling peas, washing peas, stewing peas till they are tender, frying with sugar, coagulating, cutting in pieces. Traditional recipe also requires jujube meat. Fangshan Restaurant makes the best pea flour cake.

B. Tianjin snacks

The advantageous geographical location of Tianjin has attracted visitors from all parts of the world who have brought with them various kinds of delicacies. Consequently, Tianjin has developed its unique culinary culture. Snacks of Tianjin are cheap, aromatic, tasty and attractive in color. Inhabitants of Tianjin are proud of their snacks. Among a wide variety of snacks of Tianjin, snacks of Nanshi, Hebei great alley (today's Hongqiao great alley) and Niaoshi are the most famous which are numbered at more than a hundred. Representative snacks of Tianjin include Goubuli steamed buns, Erduoyan fried rice cake and Guifaxiang 18th Street fried dough twists.

Goubuli steamed stuffed buns

Goubuli steamed buns originated during the reign of Daoguang Emperor of Qing Dynasty. It's said that the snack was created by an apprentice named Gou Zi who worked for a steaming food shop located in Wuqing District of Tianjin. Gou Zi was born when his father was already forty years old. According to Chinese tradition, a child would be easy to bring up if he or she is given a low-key name. To follow the tradition, the father gave the son the name of "Gou Zi" (meaning dog). Gou Zi started his apprenticeship at a steaming food shop at the age of fourteen. After acquiring superb skills, he opened a steamed stuffed buns shop. Since the steamed stuffed buns were very delicious, the shop turned out to be a great success. As Gou Zi was extremely busy everyday for his business, diners joked that Gou Zi was too busy to respond to our requests. Thereafter, the name of "Goubuli steamed" ("Gou Zi making no response") became widely known. After eating Goubuli steamed buns, Yuan Shikai, the then governor of Zhili, praised the food and presented the buns to Empress Dowager Cixi. After tasting the food, Empress Dowager Cixi extolled, "Animals in mountains, wild geese flying in the sky, beef and mutton, seafood from the sea — none of them is as palatable as Goubuli steamed buns. The buns can prolong life." The most distinctive feature of Goubuli steamed buns is the 15 pleats on the edge of each bun.

Goubuli steamed stuffed buns

Erduoyan fried rice cake

As one of the local snacks of Tianjin, Erduoyan fried rice cake features exquisite cooking techniques and unique style. Similar to common fried rice cake, Erduoyan fried rice cake is oblate in shape and golden yellow in color. Common fried rice cake is sticky while Erduoyan fried rice cake is crispy (hence the nickname "breaking when hitting ground"). Its fillings are tender, aromatic and sweet. Because of the unforgettable aftertaste, it has been known at home and abroad. The snack originated during the reign of Guangxu Emperor of the Qing Dynasty. A street vendor with the surname of Liu opened a fried rice cake shop along with his nephew in Erduoyan Alley. To attract more diners, he selected glutinous millet which was ground and made into skin for fried rice cake. Bean stuffing contained more brown sugar than fried rice cake made by other vendors. He also fried bean stuffing and brown sugar together (common fried rice cake does not include this step) and the resulting stuffing can be preserved for two weeks even in summer. The cake was fried in warm oil for a long time, giving the cake skin a crispy texture and unique flavor.

Guifaxiang 18th Street fried dough twists

Guifaxiang 18th Street fried dough twists is a traditional snack of Tianjin. Along with Goubuli steamed buns, Erduoyan fried rice cake, the time-honored fried dough twists are called "three treasures of Tianjin". Fried dough twists stuffed with assorted fillings which were created by Guifaxiang Company look like a golden stick dotted with rock sugar and covered with seasoning such as green & red shreds and sweetened melon strips. With an aroma of sweet osmanthus, the fried dough twists remain crispy and tasty even after being preserved for one month.

2) Style of the northeast

The northeastern region we refer to normally include the administrative regions of Heilongjiang, Jilin and Liaoning. Academically, the northeastern region also includes the northeastern part of Inner Mongolia and the northern part of Heibei. During the long history of the northeastern region, the ethnic groups living in the area since ancient times had been struggling against the hostile environment for survival. During the process, they had created a culinary culture unique to the northeastern region and thereby promoting the integration of culinary culture of the northern China and that of the southern China as well as culinary experience exchanges between the east and the west. Natural environment and fusion of ethnic groups have deeply influenced the creation of culinary culture of the northeastern region. Representative snacks of the region include baked cold noodles, glutinous steamed buns stuffed with sweetened bean paste and frozen fruits.

Baked cold noodles

Baked cold noodles, originating from Lianzhushan township of Mishan city in Jixi of Heilongjiang, is a snack widely eaten in Heilongjiang. Preparation for the snack is quite easy with eggs and sausage as the ingredients and sauce as seasoning. Baked cold noodles were first seen outside schools in Lianzhushan township in Mishan city around 1999. It was presented mainly in three forms: grilled, baked on iron plate and deep-fried. Tastes of these three versions are different with cold noodles baked on iron plate and deep-fried being the most popular nowadays. Cold noodles baked on iron plate can use eggs, sausage, dried meat floss, onion and coriander as ingredients. After being baked on iron plate, the noodles can be topped with sauce. Tender, tasty and aromatic, the noodles are chewy, making it popular snack among the common people.

Glutinous steamed buns stuffed with sweetened bean paste

As a traditional food item created by the Manchus living in northern China, glutinous steamed buns stuffed with sweetened bean paste have a history of more than 1,000 years. It has remained a popular snack in northern China till today. The traditional way of making the snack uses millet as the major ingredient and employs natural fermentation cooking technique. Flavorful, tasty, nutritious and able to resist hunger and coldness, the snack has become a food item consumed in leisure time and during festivals.

Frozen fruits

Representative frozen fruits include frozen pear, frozen persimmon and sugar-coated haws on stick. The winter in the northeast and Inner Mongolia is long and cold. The outdoor space is a natural freezer. The unique natural environment provides favorable conditions for preservation and freezing of food. Having been exposed to cold climate, the northerners have developed a preference for "frozen food" and thereby created the tradition and culinary culture of consuming "frozen food". The tradition has been developed from a cold environment as well as the exploitation of the natural environment by the northerners.

Glutinous steamed buns stuffed with sweetened bean paste

3) Style of the southwest

Similar to its "mountain culture", snacks of the southwestern region created by the ethnic minority groups exhibit a strong flavor of the local style and that of the ethnic groups. Among the 55 ethnic groups in China, a large number of them live in the southwestern region. Population of ethnic minority groups in Yunnan, Guizhou and Guangxi accounts for about 31% of the total population of the three provinces. Consequently, snacks of the southwestern region have distinctive features of ethnic minority groups living in the region. Representative snacks include Dandan noodles, crossing-over bridge rice noodles and fish in sour soup.

Dandan noodles

Dandan noodles is a famous local snack in Sichuan. The noodles were sold by vendors who usually carried a load, hence the name "Dandan noodles" ("Dan" in Chinese means load). The noodles is made of flour and topped with minced pork when ready. The noodles are thin, aromatic with gravy, salty and slightly spicy.

Crossing-over bridge rice noodles

With exquisitely made soup, unique way of eating and palatable taste, crossing-over bridge rice noodles is a local snack of Yunnan. The snack consists of four components: soup, sliced meat, rice noodles and seasoning. The soup is covered with a layer of hot oil. Process of preparation: peel pigeon's eggs and put them into the bowl; put sliced meat such as fillet, sliced chicken and sliced pork into the bowl and mix them so as to cook the meat by using the heat in the bowl; put fixings and seasoning in the bowl; a bowl of delicious rice noodles with rich flavor is then made.

Crossing-over bridge rice noodles

Fish in sour soup

Fish in sour soup is a famous dish of Dong ethnic group who live in the boundary land between Guizhou, Guangxi and Hunan. Miao ethnic group, Shui ethnic group and Yao ethnic group who are neighbors of the Dong ethnic group also have a similar dish with the version of the Dong ethnic group living in Guizhou being the most famous. According to relevant research, the dish originated in Leidong township in Liping County. Raw materials for making the dish include fish, sour soup and spices such as litsea cubeba. As one of the representative dishes of Guizhou cuisine, the dish is slightly sour, aromatic, tender, tasty and able to stimulate the appetite. Preparation process: making sour soup; removing viscus of fish; cooking fish in the sour soup.

4) Style of the southeast

Snacks of the southeastern region are mainly composed of snacks of Guangdong and Fujian. Snacks of this region are usually exquisite in appearance and offered in a dazzling variety. Due to the climate and geographical location of the region, rice and seafood are the two types of raw materials often used. Representative snacks include Shunde double-layer milk custard, fried oyster and seven-star fish balls.

Shunde double-layer milk custard

Created accidentally during the late Qing Dynasty by a farmer living in Shunde of Guangdong, Shunde double-layer milk custard requires complicated preparation process. The milk custard has a white and smooth texture, rich aroma and sweet taste. It can be consumed along with red beans, lotus seeds, mango and sesame paste, creating an unforgettable culinary experience.

Fried oyster

Fried oyster is a standing dish originating in Quanzhou, Fujian. Preparation process of Fujian version of fried oyster: placing several big and fresh oysters on hot iron plate; stir-frying the oysters; adding thicken soup, Chinese cabbage, bean sprouts and an egg. Preparation process of Tainan version of fried oyster: mixing sweet potato starch (with water) with oyster, pork and mushroom; deep-frying the mixture and making it into a cake. In September 2018, fried oyster was selected as one of the ten classical dishes of Fujian.

Seven-star fish balls

Created at the early Qing Dynasty, seven-star fish balls are fish ball stuffed with fillings. Preparation process: mixing minced fish meat with potato starch; using pork as fillings and making the mixture into balls; boiling the balls in soup; the balls float in the soup looking like stars in the sky, hence the name "seven-star fish balls". "Seven-star fish balls" is a famous snack in Fujian.

5) Style of region south of the Yangtze River

Snacks of region south of the Yangtze River are primarily composed of dim sum. Since rice and wheaten food from Jiangsu are the most famous among those produced in Jiangsu, Zhejiang and Shanghai which are located in the lower reaches of the Yangtze River, dim sum from this region is called Jiangsu-style dim sum or "dim sum of Jiangnan" (dim sum of region south of the Yangtze River). Since the region south of the Yangtze River is a land of plenty, the prosperous economy and abundant produce have promoted the development of its culinary culture, especially its creation of different kinds of dim sum. Well-known dim sum in this region includes crab soup dumpling, tang yuan and zong zi.

Crab soup dumpling of Zhenjiang

Crab soup dumpling of Zhenjiang, commonly known as crab dumpling, is a traditional snack of Zhenjiang. With a history or more than 200 years, the snack is popular in area around Shanghai and Nanjing. It has also enjoyed fame at home and abroad. With crab oil and pork as the main ingredients, crab soup dumpling is made with exquisite techniques. Delicate in appearance, it looks like a clock when placed on a steamer and a lantern when held by chopsticks. The dumpling has thin skin and plenty of soup and ingredient. If accompanied by black rice vinegar of Zhenjiang and shredded ginger, the flavorful taste would be greatly enhanced and the greasy feeling would be reduced.

Ningbo tang yuan

Tang yuan is a famous traditional snack in Ningbo of Zhejiang. With a long history, it's also one of the representative snacks in China. It is said that tang yuan was first created during the Song Dynasty. Back then, a new type of snack became popular in Ningbo. It used confectionery as fillings which were wrapped with glutinous rice flour and then made into balls. After being boiled, the glutinous rice balls were tasty and aromatic. The preparation process was also very interesting as the glutinous rice balls would sink or float in the wok when boiled, hence its earliest name of "floating yuanzi". People of other regions then called the snack "yuan xiao". Unlike the northerners, people of Ningbo have the tradition of eating tang yuan together with family members on the morning of the Spring Festival.

Wufangzhai zongzi of Jiaxing

Created in 1938, Wufangzhai zongzi has a history of more than 80 years. Dubbed as the "King of Jiangnan zongzi", Wufangzhai zongzi is known for the following features: moderately glutinous, oily yet not greasy, tender and aromatic meat, well-balanced salty and sweet taste. Elaborately made by using traditional

Wufangzhai zongzi

techniques, Wufangzhai zong zi uses high-quality raw materials. Zongzi stuffed with pork uses first-rate white glutinous rice and lean ham. Sweet zongzi uses "Da Hongpao", top grade red beans. Zongzi is made through such steps as mixing ingredients, seasoning, wrapping and boiling. Wufangzhai zongzi of Jiaxing is presented in dozens of varieties such as zongzi stuffed with pork, zongzi stuffed with sweetened bean paste and yolk.

6) Style of Hubei and Hunan

Snacks of Hubei and Hunan are the representatives of snacks of the central southern China. As the locals have a preference for spicy food, the snacks are usually spicy. Some snacks are salty or sweet able to stimulate appetite. Representative snacks include hot and dry noodles, stinky tofu and tasty lobster.

Wuhan hot and dry noodles

Wuhan hot and dry noodles, Shanxi sliced noodles, Yifu noodles of Guangdong and Guangxi, Sichuan noodles with peppery sauce and fried bean-paste noodles of the North are the five most well-known noodles in China. Preparation process of Wuhan hot and dry noodles: boiling noodles in water; drenching with cold water and oil; add sauce made from sesame soy, sesame oil, black rice vinegar and chili oil; topping with ingredients such as dried sea shrimp and spiced pickles to enhance the flavor. The noodles are smooth and chewy with rich aroma of the sauce, a delicacy whetting the appetite.

Stinky tofu

People call stinky tofu "stinky dried bean curd". With black color and smelly odor, stinky tofu of Changsha is different than those of the other places. Stinky tofu of Changsha smells bad yet tastes good. Crisp outside and tender inside, stinky tofu has a strong spicy aroma, making it an irresistible delicacy.

Tasty lobster

As a local snack of Hunan, tasty lobster is made from small lobsters. Numbing,

Stinky tofu

spicy and flavorful, it's an ideal food item goes with beer for a summer evening meal. When night falls in summer, diners eating tasty lobsters can be seen everywhere in Changsha. Although their cheeks turn red because of the spicy taste, their eyes damp with tears and head is covered with sweat, they still indulge themselves in enjoyment of the delicacy.

7) Style of the northwest

Snacks of Shanxi and Shaanxi are the representatives of snacks of the northwest. Due to local people's preference for wheaten food, their snacks are mostly made from flour. The snacks usually have rich taste and exhibit distinctive local features. Representative snacks include pita bread soaked in lamb soup, Shaanxi cold noodles and cat ear.

Pita bread soaked in lamb soup

Pita bread soaked in lamb soup is the most famous and distinctive snack of

Xi'an. White flour scones are broken into pieces and soaked in lamb soup, making a dish with palatable soup and pita bread. The snack requires exquisite techniques to make. With strong flavor, tender meat and thick soup, the snack is nutritious and oily yet not greasy. Capable of warming stomach and resisting hunger, the snack has remained popular with people in Xi'an and the northwestern region. Among the large number of restaurants specializing in pita bread soaked in lamb soup, the time-honored restaurants of "Lao Sun Jia" and "Tong Sheng Xiang" are the most well-known.

Pita bread soaked in lamb soup

Shaanxi cold noodles

Shaanxi cold noodles consist of two categories: rice-based and millet-bases. As rice-based version is the most popular, the snack of Shaanxi cold noodles is also called rice noodles. Preparation process: cutting the dough into 0.5 centimeter wide strips; adding shredded cucumber as an auxiliary; adding salt, vinegar, soy sauce, sesame sauce and chili oil and mixing all the components. Tasty with the aroma of vinegar, the noodles are refreshing in taste capable of relieving heat in summer.

Cat ear

Cat eat is a local snack of Shanxi. Process of making cat ear: cutting kneaded dough into pieces, each of the size of a horsebean; pinching the dough piece with the thumb and the index finger and turning it into the shape of a cat ear; boiling in water and taking out; mixing with leek, shredded pork, shrimp meat, dried mushroom and ham and stir-frying the mixture over high heat. The finished cat ears are soaked with soup making the snack a palatable delicacy.

8) Style of Shanghai

As the leading city of the Yangtze River economic zone, Shanghai is a world top financial hub and one of the largest metropolises in the world in terms of population and land area. Due to these strengths, Shanghai has attracted snacks from home and abroad which have been enhanced with local flavor. In Chapter Eight (titled "Metropolis Traditions Combining Chinese and Western Elements") of Volume Five (titled "Society of the Late Qing Dynasty") of *General History of Shanghai*, there is a description of the flourishing catering industry and new styles of culinary experience in Shanghai. As analyzed in the book, since Shanghai was opened up to trade, "the goal of its culinary culture has shifted from meeting basic needs and leisure purposes to social interaction and entertainment. People started to pursue high-grade food and various tastes." The book also pointed out that "the shift was the result of the Taiping Rebellion. The consequent increase in population in the concession and floating population meant more demands for food. Arrival of immigrants from other parts of the country had led to demands for different styles of food which diversified the culinary structure of Shanghai." Due to these historical factors, snacks of Shanghai exhibit features of snacks of other areas and at the same time, they have retained Shanghai traditional style. Representative snacks of Shanghai include pan-fried pork bun, small steamed stuffed bun, pork chops with rice cakes and Four Guardian Warriors.

Pan-fried pork bun

Also known as "pan-fried steamed stuffed bun" or "pan-fried steamed bun", pan-fried pork bun is a traditional snack popular in Shanghai, Zhejiang, Jiangsu, and Guangdong. Materials used include flour, sesame, chopped green onion, pork and pigskin jelly. It used to be sold at teahouse and hot water store as a sideline item. The fillings are mostly composed of pork and pigskin jelly. Since the 1930s, stores exclusively devoted to pan-fried pork bun had been opened in Shanghai. New types of fillings were created such as corn, shepherd's purse and shrimp meat. Today, "Yang's Fried Dumpling" restaurant is well-known throughout the country.

Pan-fried pork bun

Nanxiang small steamed stuffed bun

Originally called "Nanxiang big meat bun", "Nanxiang big steamed bun", "Guyi Garden small steamed stuffed bun", "Nanxiang small meat bun" and "Nanxiang small steamed bun", Nanxiang small steamed stuffed bun is a traditional snack of Nanxiang township in Jiading district of Shanghai. The snack is known for its thin skin, tender meat, plenty of gravy, palatable taste and appealing appearance. To make the fillings, ham is first made into meat paste, and then small amount of bruised ginger is added along with pigskin jelly, salt, soy sauce, sugar and water. The wrapper is made from refined flour which has not fermented. In August 2014, according to the list of the recommended projects for the list of the 4th batch of representative projects for national-level intangible cultural heritage released by the Ministry of Culture, technique for making Nanxiang small steamed stuffed bun was listed as national-level traditional technique for making wheaten food.

Pork chops with rice cakes

Pork chops with rice cakes is a cheap and unique traditional snack of Shanghai which has a history of more than 50 years. Pork chops and thin and small rice cakes are quick-boiled in oil and then braised. The snack has both the rich aroma of pork chop and the glutinous and crispy taste of rice cake, making it a palatable food item. Covered with golden luster, pork chops are crispy outside and tender inside. The snack is glutinous, aromatic, slightly sweet-and-spicy and flavorful.

Four Guardian Warriors

A typical breakfast in Shanghai includes the following "Four Guardian Warriors": Chinese pancake, deep-fried dough sticks, soybean milk and glutinous rice ball.

Chinese pancake is either sweet or salty. Sweet pancake uses white sugar as fillings and is covered with white sesame. The pancake is aromatic, sweet, and flavorful, releases burnt scent. Salty pancake is made with chopped green onion, a technique requiring high-level skills.

Deep-fried dough sticks are famous in Shanghai as well as the rest of the country. People in Shanghai usually wrap deep-fried dough sticks with Chinese pancake or accompany deep-fried dough sticks with soybean milk and glutinous rice ball.

Unlike the northern region where soybean milk is usually sweet, soybean milk in Shanghai is either sweet or salty. Sweet soybean milk is made by adding white sugar. Salty soybean is made by adding dried small shrimps, deep-fried dough sticks, pickled mustard tuber, light soy sauce or specially made sauce and sesame oil.

Glutinous rice ball mainly uses glutinous rice which is stuffed with deep-fried dough sticks, pickled mustard tuber and dried meat floss. The snack is aromatic and chewy. Sugar or black sesame had once been used as stuffing which was then replaced by pickled mustard tuber and dried meat floss.

2. Classification by raw materials

Geographical environment can significantly influence lifestyle and production mode of residents of a specific region, which in turn influence their culinary traditions. China is a country with a vast territory and huge differences exist among regions in terms of terrain and climate. As a result, crops grown and livestock raised in each region are different. Therefore, if based on by raw materials used, snacks can be classified into different groups. This chapter takes a brief look at four groups of snack: rice; wheat; meat; seafood.

1) Rice

According to modern day archaeological study, the ancestors of inhabitants living in coastal area of southeastern China started to cultivate rice for survival during the Neolithic period. With migration of population, rice-based culture (rice culture) fused with dry land culture (millet culture) which had originated in the Yellow River basin in the North. Over time, the splendid Chinese civilization had been created. With the progress of history, the status of rice in China and the corresponding rice culture had been gradually improved. Millet, wheat and rice were the three major crops in ancient times. From the late Neolithic period to the Shang Dynasty and Zhou Dynasty period, millet was the leading crop followed successively by wheat and rice. Since the Qin Dynasty and Han Dynasty, wheat almost caught up with millet in terms of status while rice was still ranked the third. From the Three Kingdoms period to the Wei, Jin, Southern and Northern Dynasties period (3rd century to 6th century), rice became as important as millet and wheat due to the development of the southern region. Since the Tang Dynasty and Song Dynasty (7th century to 13th century), rice has replaced millet and wheat as the No. 1 crop.

2) Wheat

Since ancient times, the Chinese people have a tradition of making grains and starch into strip-shaped food. According to findings of archaeological excavation, Lajia noodles of Qinghai dating back 4,000 years are the earliest noodles invented. The noodles were made from millet and broomcorn millet. In early times, wheat and barley were eaten in grains, the way rice was eaten. Wheat started to be made into flour during Han Dynasty leading to sharp increase in variety of wheaten food. The primitive forms of steamed bun, pancake, noodles, steamed stuffed bun and jiaozi appeared during the period and flour fermentation technology was invented. As a result, a large number of snacks using wheat grains (mainly wheat) as raw material have been created across the country.

3) Meat

In ancient times, the Chinese people "ate fruits of the trees and flesh of birds and animals; they ate animal fleas raw and drank its blood." They just lived the life of a savage. Legend has it that Sui-Ren Shi found out how to start a fire and taught the ancient people to cook food over fire in order to remove raw taste, hence the beginning of "cooking". As a result, variety and scope of food suitable for consumption had been increased. Development of meat consumption culture in China had been closely linked to improvement of cooking techniques and use of fire. Other factors having contributed to the rapid development of meat consumption culture in

China include: a vast territory of the country; complex and diverse climatic and geographical environment; robust livestock farming; mutual influence among culinary cultures of different ethnic groups. Snacks using "meat" as component have been created in large numbers such as lamb skewer of Xinjiang, Nanxiang small steamed stuffed bun and pork chops with rice cakes.

4) Seafood

China has a continental coastline extending for 18,000 kilometers. If adding the coastlines of all the islands, it would be 32,000 kilometers. The vast coastline provides a wide variety of seafood for Chinese culinary culture. Snacks with seafood as component have also appeared in a large number such as sate noodles and seven-star fish balls of Xiamen in Fujian.

Chapter 5 Tea Culture & Wine Culture

5.1 Tea Drinking

1. Birthplace of tea

In 1824, Major Robert Bruce, a British official stationed in India, found a wild tea tree in Sadiya, Assam of India. The tree was 13.1 meters high with a diameter of approximately 0.9 meters. Hence, some scholars from the western world believe that India is the birthplace of tea. Rather, China is the real birthplace of tea. As recorded in *The Classic of Tea*①, a book from the Tang Dynasty, tea tree with a diameter of about 1 meter had already been found during the Tang Dynasty. Its height was almost the same as that of the tea tree found by the British major, but it was found 1,100 years earlier than the Indian tea tree.②

Tea has a history of around 4,000 years in China. During the Tang Dynasty (618 - 907), Japanese monks introduced tea into Japan and combined it with Zen creating the world renowned Japanese Tea Ceremony. During the 17th century, the Dutch people brought tea drinking practice to Europe which then spread to the entire

① *The Classic of Tea* is the first monograph on tea in the world, written by Lu Yu, the creator of Chinese tea ceremony.

② Yu Ye. *Chinese Tea (Chinese and English Version)* [M]. Hefei: Huangshan Publishing House, 2011: 2 - 4.

European continent. In Britain, it had developed into the culture of drinking afternoon tea. Prior to the 19th century, all the drinkable tea leaves in the world were produced in China.① It was the Chinese people who first discovered, cultivated and drank tea. They also introduced tea plant seeds, tea drinking methods and tea cultivation techniques to the rest of the world. Tea is one of the major contributions the Chinese people have made to the world culture.

2. Origin of the Chinese character "茶" (cha)

The Chinese character for "茶" (cha) was written as "荼" (tu) before the Tang Dynasty. "荼" (tu) is a kind of sowthistle herb grown in ancient times. As a phonogram, the radical of "荼", "艹", represents the meaning of the word signifying that "荼" is an herb. However, it's later found that tea is a woody plant rather than an herb. Therefore, it's incorrect to use the character of "荼" (tu) to refer to tea.

The Chinese character of "荼" (tu) was simplified into the character of "茶" (cha) during the Han Dynasty. The additional horizontal stroke in "荼" was deleted, hence the character of "茶". During the mid-Tang Dynasty, the pronunciation was changed from "tu" into "cha", thus the pronunciation, writing and meaning of the character for tea had been gradually finalized. Due to the popularity of *The Classic of Tea*, an encyclopedia on tea written by Lu Yu, more and more people started to drink tea and at the same time, use the character of "茶" to refer to tea. Hence, the writing of the character of "茶" was further finalized and is still used today.②

3. Chinese people and tea drinking traditions

For the Chinese people, there are seven necessities for daily life: firewood,

---

① Ye Lang and Zhu Liangzhi (author), Cathy (translator). *Chinese Culture Reader (German Version) (Blick auf die Chinesische Kultur)* [M]. Beijing: Foreign Language Teaching and Research Press, 2014: 225.

② Yu Ye. *Chinese Tea (Chinese and English Version)* [M]. Hefei: Huangshan Publishing House, 2011: 9.

rice, oil, salt, sauce, vinegar and tea. Tea is the only beverage among the seven, yet it's as important as food. The Cantonese usually go to the teahouse for breakfast. They would order a pot of tea and several dishes of dim sum before going for work. For the Chinese people, although tea cannot satiate hunger, it's a beverage they must drink on a regular basis. Drinking tea can not only quench thirst, it's also a way of life. Instead of meeting people's physical needs, tea drinking is a way for people to seek spiritual value.①

The Chinese people drink tea at any place and at any time. Lin Yutang, a Chinese linguist, once said that Chinese people have a passion for tea drinking. They drink tea at home and at meeting. When they mediate between quarreling parties, they drink tea. They drink tea at breakfast and after lunch. When having a guest visiting home, the host would prepare a pot of tea to entertain the guest, which is a common way of entertaining guests. At a banquet, the host would serve wine to guests to express respect and tea is the only beverage which can replace wine. "Replacing wine with tea" is a widely acceptable practice which follows etiquette rules. Tea is a beverage consumed in daily life of Chinese people. It's also a must-have for banquet, wedding party, birthday party, funeral and worshiping ceremony.②

Tea house culture is popular in many cities in China. Similar to cafe in the western world, Chinese teahouse is a place for leisure and entertainment. However, their overall ambience is different: teahouse is filled with excitement while cafe is usually quiet. The excitement of a teahouse comes from the entertainments offered there such as Quyi performance, singing & telling performance and Pingshu. People watch the performance while drinking tea, just enjoying a leisure time. Modern-day teahouse serves various social functions. It can be a place for drinking tea, having a leisure time, engaging in business negotiation and exploring tea culture.③

---

① Yu Ye. *Chinese Tea (Chinese and English Version)* [M]. Hefei: Huangshan Publishing House, 2011: 29.

② Yu Ye. *Chinese Tea (Chinese and English Version)* [M]. Hefei: Huangshan Publishing House, 2011: 29 – 31.

③ Yu Ye. *Chinese Tea (Chinese and English Version)* [M]. Hefei: Huangshan Publishing House, 2011: 32 – 33.

According to the theory of traditional Chinese medicine, tea can cure many diseases due to the fact that it's slightly bitter in taste, cool in nature and contains components which have curative effects. Modern pharmacology has corroborated the theory. The following lists the 14 benefits of drinking tea:

(1) Invigorating; enhancing memory and thinking capability.

(2) Reducing fatigue; promoting metabolism; maintaining the normal functions of heart, blood vessel, stomach and intestines.

(3) Preventing tooth decay; according to a research done in the UK, tooth decay may be reduced by 60% among children if they regularly drink tea.

(4) Tea leaves contain large quantities of microelements which are beneficial for health.

(5) Tea leaves can inhibit malignant tumor growth; drinking tea can effectively inhibit the growth of cancer cells.

(6) Inhibiting cell aging and prolonging life; anti-aging effect of tea leaves is more than 18 times that of vitamin E.

(7) Delaying and preventing formation of intimal lipid plaque; preventing arteriosclerosis, hypertension and cerebral thrombosis.

(8) Activating central nervous system and increasing athletic stamina.

(9) Effective in weight loss and having cosmetic effects (oolong tea in particular).

(10) Preventing senile cataract.

(11) Tannic acid contained in tea leaves can kill a variety of bacteria and consequently prevent stomatitis, laryngopharyngitis and diseases easily contracted in summer such as enteritis and dysentery.

(12) Drinking tea can protect hematopoietic functions. As tea leaves contain radiation-proof materials, drinking tea while watching TV can reduce radiation and protect eyesight.

(13) Drinking tea can help maintain a normal acid-base balance in the blood. Tea leaves contain alkaloids such as theine, theophylline, theobromine and

xanthine making tea an ideal alkaline beverage.

(14) Preventing heatstroke and relieving summer heat. Skin temperature would drop 1 – 2 degrees Celsius 9 minutes after drinking hot tea giving a cool and dry feeling.①

For tea drinkers, the first feature of tea drinking is "purifying body". Tea trees usually grow in very clean environment. The cleaner the environment, the better the quality of tea leaves. High-grade tea trees are usually found on high mountains and therefore surrounded by clouds in fresh and clean air. Tea leaves of this type of tree are tender and even covered with dew when picked exuding a naturally fresh aroma. Tea made from high-quality tea leaves is clear in color and has pleasant smell. It can purify the body and harmonize body functions.

The second feature of tea drinking is "seeking peace of mind". Confusion, conflict, boredom and tension we experience in this complicated world would do harm to us both physically and mentally. A cup of tea may allow us to temporarily escape from the noise and chaos of the world. Sipping the tea, we may have peace mind as if we're sitting beside a lake on a quiet night and the bright moon shines over the world. With a cup of tea, we may easily find a path leading to a new world.

The third feature of tea drinking is "expressing respect". According to an old tradition of China, serving tea for a guest is a way of showing respect. The guest also expresses his or her respect by receiving the tea and drinking, whether he or she is thirsty or not. After drinking the tea, the guest may feel invigorated. At a banquet, those who are late would use tea to replace wine and drink several cups of tea in a row as a punishment. In some regions in China, three cups of tea are served for a guest signifying a warm welcome, hospitality and best wishes respectively. Pouring

---

① Li Lin. *Introduction to Tea and Wine Culture*[M]. Taiyuan: Beiyue Literature and Art Publishing House, 2010: 47 – 48.

tea for guest is an act of showing both hospitality and respect.[①] There is also another tradition in China where the guest "picks up a cup of tea in order to ask the guest to leave". By pretending to drink tea, the host is politely asking the guest to leave.

4. Six categories of tea

Currently, there is no unified approach available to classify tea. On the basis of process used and color of tea, tea can be classified into six categories: green tea; black tea; dark tea; Oolong tea; yellow tea; white tea.

1) Green tea

Green tea is tea not fermented. After drying, fresh tea leaves are stir-fried in hot pot with a temperature of 100 to 200 degrees Celsius so as to ensure the tea retains its green color. Processing steps usually include fixation, rolling and drying. Based on different drying methods used, green tea can be classified into fried green tea, baked green tea, steaming green tea and sun-cured green tea.

As a type of tea with the largest output in China, green tea is produced by almost all tea production areas in the country. China also boasts the widest variety of green tea in the world. Due to its enticing aroma, rich taste, attractive appearance and suitability for several rounds of brewing, green tea is popular among both domestic and foreign consumers. Famous varieties of green tea are: Xihu Longjing tea; Dongting Mountain Biluochun tea; Lushan Yunwu tea; Huangshan Maofeng tea; Taiping Houkui tea; Mount Emei Zhuyeqing tea; Mengding Ganlu tea; Xinyang Maojian tea; Guzhu Zisun tea; Putuo Buddha tea; Duyun Maojian tea; Nanjing Yuhua tea.[②]

---

① Ye Lang and Zhu Liangzhi (author), Cathy (translator). *Chinese Culture Reader (German Version) (Blick auf die Chinesische Kultur)* [M]. Beijing: Foreign Language Teaching and Research Press, 2014: 227 - 228.

② Yu Ye. *Chinese Tea (Chinese and English Version)* [M]. Hefei: Huangshan Publishing House, 2011: 44 - 45.

2) Black tea

Unlike green tea, black tea is fully fermented tea. Due to its prominent features of red liquid and red leaves, this type of tea is called black tea. Famous varieties include Qi Hong, Dian Hong and Ning Hong.

One major difference between black tea and green tea is the processing method. Black tea processing steps do not involve fixation. Instead, it undergoes withering so that the fresh leaves would lose a portion of water contained. Subsequent steps include: rolling (rolling into strips or cutting into particles); fermentation (oxidizing tea-polyphenol and changing it into red chemical compounds). Some of these chemical compounds dissolve in water while the remaining compounds do not dissolve in water and accumulate in blade creating red liquid and red leaves. Varieties of black tea include Souchong black tea, Congou black tea and broken black tea.[①]

3) Dark tea

Dark tea is fermented tea and unique to China. Raw materials used for making dark tea are coarse. Processing process involves long duration of accumulation and fermentation in order to create dust-color leaves and black liquid.[②] Dark tea is a daily necessity for ethnic groups living in border areas of China such as the Tibetan people, the Mongolian people and the Uyghurs. Dark tea such as Yunnan Pu'er tea was once sold in border areas. Famous varieties include Hunan Heimao tea, Hubei Laoqing tea, Western Route tea and Southern Route tea of Sichuan and Guangxi Liu Bao tea.[③]

Pu'er tea is made by pouring water on cured green tea and undergoing fermentation process. Health benefits of Pu'er tea include anti-aging, lipid lowering,

---

① Yu Ye. *Chinese Tea (Chinese and English Version)* [M]. Hefei: Huangshan Publishing House, 2011: 88-89.

② Yu Ye. *Chinese Tea (Chinese and English Version)* [M]. Hefei: Huangshan Publishing House, 2011: 80-81.

③ Qin Dadong. A Brief History of Dark Tea [J]. *Journal of Tea Business*, Issue 6, 1983: 33.

weight loss and blood pressure reduction. It's popular in the Southeast Asia and Japan. However, Oolong tea is more effective in reducing weight.

4) Oolong tea

Oolong tea is partially fermented tea with characteristics of both black tea and green tea. It has the freshness and strong flavor of green tea and the sweetness of black tea. Among the six categories of tea, the processing and brewing steps of Oolong tea are the most complicated. Therefore, drinking Oolong tea is also called drinking Congou tea.

Partially fermented tea refers to the tea which is moderately fermented when processing so that leaf blades would slightly turn red. Since the middle part of leaf blade is green and the edge is red, leaves of Oolong tea are described as "green leaf and red edge". In addition to health benefits common to other varieties of tea, Oolong tea can also prevent arteriosclerosis, reduce weight and promote body building. Famous varieties include Wuyi Yan tea, Anxi Tieguanyin tea, Fenghuang Dangcong tea and Taiwan Dong Ding Oolong tea.

5) Yellow tea

As slightly fermented tea, yellow tea is featured with yellow leaves, yellow liquid, pleasing aroma, rich flavor and refreshing taste. The well-known Junshan Yinzhen tea is yellow tea. Yellow tea is similar to green tea in production method except that yellow tea requires three-day heaping for yellowing step.

Before or after rolling, or before or after preliminary drying, yellow tea undergoes heaping for yellowing process and thereby creating yellow leaves and yellow liquid. Based on the tenderness and size of tea leaves, yellow tea can be classified into the following categories: "yellow bud tea" (such as Junshan Yinya, Mengding Huangya tea and Mogan Huangya tea); "yellow small leaf tea" (such as Beigang Maojian tea, Weishan Maojian tea, Yuan'an Luyuan tea); "yellow large leaf tea" (such as Huoshan yellow large leaf tea and

Guangdong Dayeqing tea).[①]

6) White tea

White tea is slightly fermented tea which is covered with white hairs, hence the name "white tea". It's mainly produced in counties in Fujian including Fuding, Zhenghe, Songxi and Jianyang. After brewing, tea leaves loosen and remain whole. Tea liquid is pale in color or colorless for first round of brewing with rich aroma.

White tea originated in China. Due to its simple processing steps, people usually say "the lazy people love making white tea." Its processing does not involve stir-frying and rolling steps. Just drying the tender tea leaves with hairs on the back of leaf blade in the sun or drying over a gentle heat so that the white hairs remain. Varieties of white tea include Yinzhen, Baimudan, Gongmei and Shoumei.[②]

5. Ten most famous teas

Famous teas of China enjoy a high reputation in the world. Although there is no universally agreed definition of famous tea, a famous tea must have its unique features if judged from four aspects of its tea leaves including color, smell, taste and appearance. There are many version of list of "ten most famous teas" in China. Although some famous teas may fail to meet all the judging criteria, they are still renowned for certain distinctive features. Based on criteria for selection of national "ten most famous teas" in 1959, this chapter introduces characteristics and identification methods for each famous tea.

1) Xihu Longjing tea

As the most famous green tea, Xihu Longjing is also the No.1 tea in China. It's

---

① Yu Ye. *Chinese Tea (Chinese and English Version)* [M]. Hefei: Huangshan Publishing House, 2011: 72-73.
② Yu Ye. *Chinese Tea (Chinese and English Version)* [M]. Hefei: Huangshan Publishing House, 2011: 87.

named after Longjing, the place in the West Lake mountainous area of Hangzhou where the tea is produced. Situated at the foot of hill and beside lake, the place has a mild climate and is shrouded in mist all year round creating abundant rainfall. Besides, due to the loose structure and fertility of soil, the tea trees usually have deep roots and luxuriant leaves and remain green throughout the year. Longjing tea is renowned for its four prominent features: green color, rich fragrance, smooth taste and attractive appearance. Tea leaf is delicate and flat in shape with uniform size and appearance. One bud produces one leaf or two leaves. Tea leaf is yellowish green, similar to the color of coarse rice. Tea liquid is aromatic with smooth taste and green color. Fake Longjing usually has the smell of grass and many old leaves. Fake Longjing tea leaves also feel somewhat coarse.

2) Dongting Biluochun tea

As another famous variety of green tea, Biluochun is grown in the Dongting mountain region near Lake Tai, Wu County, Suzhou, Jiangsu Province. It was designated as tribute tea during the Tang Dynasty. For high-grade Biluochun, one kilogram of dry tea requires 136,000 to 150,000 buds. After stir-fried, the dried tea leaves are compact with white hairs and silver green color. Since tea leaf is rolled into the shape of snail meat and the tea is cropped in spring, hence the name "Biluochun". The tea is usually brewed in porcelain cup. After infusion, white foams roll in the cup emitting strong aroma. The tea liquid has a unique flora fragrance. Bud of fake Biluochun has two leaves. Bud leaves are yellow in color and not uniform in length.

3) Huangshan Maofeng tea

As a renowned variety of green tea, Huangshan Maofeng grows in Huangshan region of Anhui. Due to the high mountains, dense forests, short sunshine duration and frequent presence of cloud and mist of the region, high-quality Maofeng tea is easily cultivated. As the newly made tea leaves are covered with white hair, the buds are sharp in appearance and leaves are picked from the peaks of Huangshan, this type of tea is called "Huangshan Maofeng". Picking and production of Huangshan Maofeng involves meticulous work. Picking takes place during the period from Qingming to Summer Begins. To ensure freshness, if picking takes place in the morning, production is then conducted in the afternoon. If picking is done in the afternoon, production is then finished at night. Maofeng tea leaves are thin, flat and slightly curly in appearance resembling bird's tongue. The fragrance resembles that of white orchid, the taste is smooth and the aftertaste is long-lasting. Fake Maofeng tea is khaki in color with bitter taste, and tea leaves are not uniform in size.

4) Lushan Yunwu tea

As a traditional tea consumed mainly by the Han people, Lushan Yunwu tea is a type of green tea. Originally a type of wild tea, it was later transformed into homegrown tea by Huiyuan, a famous monk of Donglin Temple. It started to be grown during the Han Dynasty and was designated as "tribute tea" during the Song Dynasty. Since the tea is produced in Lushan mountain region of Jiujiang in Jiangxi, it's commonly referred to as "Lushan Yunwu tea". As the major production areas are located at 800 meters above the sea level, the temperature tends to heat up slowly and therefore tea trees grown in this area mature after Grain Rain. Due to the cold and misty climate and short sunshine duration of the Lushan mountain area, Yunwu tea leaves are thick covered with hairs and are suitable for several rounds of infusion. Tea liquid is pure in taste, clear in color and release the aroma of orchid, Chinese chestnut and soybean flower.

5) Liu'an Melon Seed tea

Liu'an Melon Seed tea is produced in the Dabie Mountains area of Liu'an, Anhui. The processed tea leaves resemble melon seeds, hence the name "Liu'an Melon Seed tea". As a famous green tea, Liu'an Melon Seed tea is the only tea among the famous tea varieties whose leaves have no buds and stalks and the tea is made by pan-frying single blades of fresh leaves. This processing method constitutes a unique feature of Liu'an Melon Seed tea. Leaves picked before Grain Rain are called "carrying leaves" which are of the best quality. Leaves picked after Grain Rain are called "melon leaves". Leaves picked during plum rain season area called "plum leaves" which are slightly coarse and of mediocre quality. Only the second and third leaves are picked and the "mature" leaves rather than the "tender" leaves are selected. Tea liquid is green with sweet aftertaste. After brewing, the leaves are thick and bright. Liquid of fake tea leaves is bitter in taste and yellow in color.

6) Junshan Yinzhen tea

As a slightly fermented tea, Junshan Yinzhen is a famous variety among yellow teas in China. It's produced in Junshan Island of Dongting Lake in Yueyang of Hunan and its leaf is as thin as a needle, hence the name "Junshan Yinzhen". The practice of enjoy Junshan Yinzhen tea has high appreciation value. Transparent glass is recommended for brewing the tea as you may watch how the tea buds unfold slowly after being steeped in hot water. The bud tip turns upward and the buds suspend in midair in upright position. Tea leaves rise and fall in water and, after three rounds of "rise and fall", finally sink to the bottom of the glass slowly and wave in the water. Tea liquid is yellow in color with sweet aftertaste. Fake Yinzhen tea leaves have the smell of grass and usually fail to remain in upright position after brewing.

7) Xinyang Maojian tea

As one of the famous specialties of Henan Province, Xinyang Maojian tea is a renowned green tea sold on the domestic market in China. As it's mainly produced in Dabie Mountain areas in Xinyang and the top five tea houses in tea district produce high-quality Benshan Maojian tea, this type of tea is officially named "Xinyang Maojian". Distinctive features of leaves of Xinyang Maojian tea are: slender, round, smooth, straight, with abundant white hairs, aromatic, rich in taste and tea liquid is green in color. The tea has an enduring aftertaste. After four to five rounds of brewing, the tea liquid still retains the fragrance of ripe chestnut. Tea liquid of low-quality Xinyang Maojian is deep-green or yellow in color and muddy and has no fragrance. Tea leaves can't be brewed for several rounds.

8) Wuyi Yancha

As a famous variety of Oolong tea, Wuyi Yancha is produced in Wuyi Mountains (the most beautiful mountain in the southeast) area in the north of Fujian. Tea plants usually grow in rock crevices. As a semi-fermented tea, Wuyi Yancha has the fragrance of green tea and the smoothness of black tea making it the best variety of Oolong tea. During the 17th century, the Dutch East India Company purchased Wuyi Yancha for the first time and sold the tea through Java to all parts of Europe. Wuyi Yancha then became a daily drink of the Europeans. The tea is usually brewed in small teapot and small cup. Due to its rich aroma, the tea leaves still retain the fragrance after five to six rounds of brewing. Tea liquid is clear with bright orange-yellow or yellow color. As for the tea leaves, bottom of tea leaf is soft and bright, edge of tea leaf is red in color and center of tea leave is light green with yellow. Wuyi Yancha is warm and gentle in nature. Tea leaves can be stored for a long time without deterioration.

9) Anxi Tieguanyin tea

As a famous variety of Oolong tea in China, Tieguanying tea is mainly produced in Anxi of Fujian Province. Surrounded by hills and shrouded in mist, the average annual temperature in this area is 15 to 18 degrees Celsius. There are many versions of story about the origin of the name of "Tieguanyin" with one version matching the characteristics of the tea. Legend has it that when watching the tea leaves closely, Qianlong Emperor found the leaf solid in structure, heavy like iron, fragrant in smell and attractive in appearance, just resembling "Guanyin" (Goddess of Mercy), he then conferred the name of "Tieguanyin" to the tea. Tea liquid is clear and golden yellow in color with the aroma of orchid. The first sip is bitter and then the liquid tastes sweet. Tea leaves are suitable for several rounds of brewing. Fake tea leaf is long and thin and loses fragrance after three rounds of brewing.

10) Qimen black tea

Qimen black tea is produced in Qimen County in Anhui Province. Fresh leaves are tender and contain abundant water-soluble substances. Leaves picked in August are of the best quality making the tea a top grade black tea in China. Qimen black tea smells like both apple and orchid and the aroma is long-lasting. Top grade Qimen black tea also emits the fragrance of rose. The unique aroma of Qimen black tea is called "Qimen aroma" by many foreign consumers. Milk and sugar may be added to the tea and the tea tastes best when no refreshment is added. Tea liquid is brownish red in color. Fake tea leaves contain artificial pigments, tea liquid is bitter and lacks richness in taste, and tea leaves are not uniform in appearance.

6. Tea set

The Chinese people view tea drinking as an art of great wisdom. It seems easy to savor the taste of tea but the act of tea appreciation carries profound significance. Therefore, Chinese people give considerable attention to every aspect and element involved in tea brewing process such as water quality, temperature, tea leaves and tea set.

If water is the mother of tea, then tea set is the father of tea. In ancient China, tea set referred to tools used for tea preparation and tea drinking. In modern times, "tea set" refers to equipment used for tea drinking including tea cup, teapot, tea tray and tea spoon. Based on materials used, tea set can be classified into several categories such as ceramic, porcelain, purple sand, metal, lacquer, bamboo/wood and glass.

Made by firing potter's clay, ceramic tea set is one of the earliest examples in human history of tea set production and usage. This type of tea set was first made of coarse terrazzo, then hard pottery followed by glazed pottery, and finally purple sand pottery. Ceramic tea set was replaced by porcelain tea set at one point and regains popularity in recent years due to the flourishing of tea ceremony performance.

Porcelain tea set includes several varieties such as celadon porcelain, white porcelain, black porcelain and faience porcelain. Celadon porcelain tea set produced in Longquan of Zhejiang boasts of many advantages of other types of tea set. Additionally, liquid of green tea, if brewed by using Longquan tea set, is appealingly beautiful due to the green color of the set. However, if using Longquan tea set to brew black tea, white tea, yellow tea or dark tea, the tea liquid would lose its natural color, which is a weakness of Longquan porcelain tea set. White porcelain tea set produced in the Town of Jingde in Jiangxi Province is sold on foreign market. It can be used to brew any kind of tea leaves. Due to its exquisite design, elegant decorative patterns as well as paintings (such as mountains, rivers, four seasons, flowers, birds and animals) and celebrity calligraphy on the outside of tea utensils, Jingde white porcelain tea set has considerable artistic value and is the most commonly used tea set nowadays.

Purple sand tea set is different from ceramic tea set or porcelain tea set, but has the features of both types of tea set. Made of potter's clay, purple sand tea set contains abundant iron and is dubbed "the best mud and the best rock". Purple sand tea produced in Yixing of Jiangsu Province is of the highest quality. Fired at temperatures between 1,000 to 1,200 degrees Celsius, items of the tea set are tight in texture without any possibility of seepage. The smooth surface has a sandy feeling of fine particles. The microscopic pores of tea set items ensure absorbability and breathability. Cold-resisting and heat-resisting, the tea set is slow in heat conduction preventing hand scalding or tea set items bursting. However, it's very difficult to appreciate the graceful movements of tea leaves in water and the color of tea liquid when brewing tea with purple sand tea set.

Metal utensils including metallic tea drinking items were used during the

Southern and Northern Dynasties period in China. The manufacturing skills reached the highest level during the Sui Dynasty and Tang Dynasty. Since the Ming Dynasty, metal tea set had gradually disappeared with emergence of new varieties of tea, transformation in tea drinking methods and appearance of ceramic tea set. As tea set items made of tin, iron and lead were thought to change the taste of tea liquid if used for brewing, these metal tea set was seldom used back then. However, tea leaves storage container made of metal such as tin bottle and tin jar has been produced in large numbers and are still used today. Metal storage container has better tightness quality than container made of paper, bamboo, wood, porcelain and ceramic, and has better damp-proof and light-proof quality making it an ideal way of preserving loose tea.

First made during the Qing Dynasty, lacquer tea set is mainly produced in Fuzhou, Fujian. A wide variety of lacquer tea set has been produced in Fuzhou such as flashing sand, golden agate, glazed spun gold, simulation of antique porcelain, covered & filled in, high carving and embedded silver. With creation of new processes such as ruby-like red golden sand and veiled patterns, lacquer tea set is more beautiful and appealing.

Bamboo and wood was used to make tea set by the common people before the Sui Dynasty and Tang Dynasty. Due to such strengths as the availability of steady supply of raw materials, easy production process, zero pollution for tea and no harm done to health, bamboo/wood tea set had remained popular among people. However, as bamboo/wood tea set can't be used for a long period of time, it has no value for collection. During the Qing Dynasty, bamboo tea set was produced in Sichuan which consisted of inner casing and outer casing. Inner casing was mostly made of potter's clay and porcelain. Outer casing was made from fine neosinocalamus affinis. After several steps of processing, neosinocalamus affinis was made into soft bamboo filaments and each filament was as thin as hair. After coating and dying, the bamboo filaments were woven according to the size of the inner casing. Bamboo tea set can protect the inner casing from damage and prevent hand scalding. At the same time, bamboo tea set has great artistic value.

Bright and transparent, glass can be made into different forms and used extensively. If using glass cup for tea brewing, the color of tea liquid, tenderness of tea leaves, movement of tea leaves and unfolding of tea leaves can all be seen, creating a dynamic art appreciation appearance. If using glass tea set to brew famous varieties of tea, the vapor, clear and green liquid and gracefully unfolded tea leaves can be seen through the crystal-clear cup presenting a pleasant view to the eye. Glass tea cup is inexpensive and of high quality. However, glass tea cup is liable to crack and is more likely to scald hand than ceramic or porcelain cup does.①

Today, tea sets made of different materials are diverse in style and involve exquisite craftsmanship. Since it was first created, tea set has become increasingly delicate and exquisite and serves as an important carrier of tea culture.

Purple sand tea set

① Li Lin. *Introduction to Tea and Wine Culture*[M]. Taiyuan: Beiyue Literature and Art Publishing House, 2010: 52-70.

5.2 Wine Drinking

1. Origin of the Chinese character for "wine" ("jiu")

There is an interesting legend about the origin of the Chinese character for "wine" ("jiu"). In ancient times, a man who knew how to make wine but his wine didn't taste good. One night, he dreamed of an old fairy who told him that if he took one drop of blood from three men respectively between 5 p.m. to 7 p.m. nine days later, and put the three drops of blood in wine cellar, the wine would taste good. He followed the instructions of the old fairy by taking one drop of blood from three men respectively: a scholar who was going on a journey to take civil examinations; a general who just won a battle; a lunatic who was lying under a tree. After arriving home, he put the three drops of blood in wine cellar and the wine he finally made was tasty and aromatic. Since the old fairy told him to take the "blood of three men" "between 5 p.m. to 7 p.m.", the man then combined the radical of "氵" (symbolizing three) with the Chinese character of "酉" (meaning "between 5 p.m. to 7 p.m.") to create the Chinese character of "酒" (meaning wine). As the old fairy told him to take the wine nine days later, the pronunciation of the character of "酒" was "jiu" (the corresponding Chinese character for this pronunciation is "九" meaning "nine"). Based on this legend, wine drinkers roughly include three types of people who share the same state of mind with the above-mentioned three men respectively, i.e., the scholar, the general, the lunatic. The scholar drinks like a gentleman, the general gets very excited after three rounds of drinking and the lunatic drinks whatever is offered to him and becomes unconscious by drinking too much.

The story is just a legend. In fact, the Chinese character of "酒" first appeared as oracle-bone script which was written in two ways: simple pictograph for "酉" meaning equipment for making wine; the middle part of the character was a wine bottle with overflowing wine on both sides of the bottle emphasizing what was contained in the bottle was liquid. Having undergone the evolution from jinwen, seal script to regular script, the character of "酒" is now written as a phonogram with the character of "水" as the shape element and the character of "酉" as the phonetic element. The evolution of the writing of the Chinese character for "wine" testifies the long history of wine culture in China.[①]

2. Chinese wine and Western wine

Almost at the same time during the Neolithic period, the Chinese people used rice sprouts to make wine and the Babylonians used malt to brew beer. However, nothing is known about whether there was any correlation between them. About 3,200 years ago, the Chinese people used malt and rice sprouts as sacchariferous agents for making grain wine which was called "Li wine". Although the sweet Li wine was not called beer, it was similar to today's beer. According to historical records, the manufacturing process of making malt for Li wine is almost the same with that of making malt for today's beer. As the later generations preferred wine made out of distiller's yeast over the thin Li wine, Li wine manufacturing techniques failed to be passed down over generations. As a result, Li wine had been gradually replaced by yellow wine which is made out of distiller's yeast.[②]

Using distiller's yeast to make wine is a highlight of the wine making history of China. Thanks to the long history of the agricultural civilization of China, Chinese people used distiller's yeast to make wine more than 3,000 years earlier than the Europeans. Mold is one major microbe growing in distiller's yeast. Utilization of mold is one important invention by the Chinese people.[③] In 138 BC, Zhang Qian

---

① Yang Shuqiong. Chinese Character for "Wine" and "Wine Culture" [J]. *Journal of Inner Mongolia Radio & TV University*, Issue 6, 2017: 33-36.

② Liu Yong. *Chinese Wine (Chinese and English Version)* [M]. Hefei: Huangshan Publishing House, 2012: 56.

③ Liu Yong. *Chinese Wine (Chinese and English Version)* [M]. Hefei: Huangshan Publishing House, 2012: 66.

explored the Western Regions and brought back grapes and wine brewing artisans initiating wine brewing in China. Wine was introduced into China 700 to 800 years earlier than being introduced into France. After distillation techniques invented by the Chinese during the Eastern Han Dynasty was introduced into Europe during the 18th century, the traditional technique used by the Europeans was significantly upgraded. Prior to that, people in the Western world saccharified grain with malt starch and used microzyme rock sugar for fermentation. Another contribution the Chinese people made to the world wine making industry was the practice of boiling wine to prevent rancidity. *Beishan Classic on Wine*, compiled during the Northern Song Dynasty, detailed the techniques of boiling wine. Western countries started to employ the techniques over 700 years later.①

3. Festival wine customs

Wine played an important role in day-to-day life of the ancient people. People drank wine together to celebrate festivals or festive events. As different wine was drunk for different festival, wine had gradually become the carrier and embodiment of the messages of festivals. For example, on the first day of the first lunar month, the common people would drink "Tu Su" wine or wine infused with pepper as these two kinds of wine can dispel pestilential vapor and symbolize peace and longevity. By drinking the wine, people expressed wishes for peace and good health for all family members in the coming year. The Dragon Boat Festival falls on the fifth day of the fifth lunar month when people commemorate the great poet Qu Yuan. On this day, people drank calamus wine and realgar wine to drive away evil spirits and enhance health as the fifth lunar month was viewed as an evil month in the past. Since ancient times, people normally gazed at the moon and drank wine to celebrate the Mid-Autumn Festival which falls on the 15th day of the 8th lunar month. Drinking wine during the festival symbolizes family reunion and

① Liu Yong. *Chinese Wine (Chinese and English Version)* [M]. Hefei: Huangshan Publishing House, 2012: 1-2.

homesickness. During the Double Ninth Festival which falls on the 9th day of the 9th lunar month, the ancient people would drink chrysanthemum wine to pray for longevity.

Most of these customs are still followed today and they have been strongly influenced by the unique medicinal liquor culture of China. Wine has therapeutic effects and medicinal liquor combines wine with health-enhancing traditional Chinese medicine and food items. Therapeutic effects of the medicinal liquor are consequently improved as wine and medicine mutually promote each other in their curative effects. Wine drank during the traditional festivals such as pepper wine, cypress wine, "Tu Su" wine, calamus wine and chrysanthemum wine can all be used for disease prevention and treatment. Due to the poor medical and health conditions in ancient times, these wines had played an important role in disease prevention by the common people.[①]

4. Top ten liquor brands

Since ancient times, Chinese liquor has played a significant role in the life of Chinese people and serves as an indispensable part of various occasions such as socializing, festive events and gift giving. Chinese famous liquor brands are selected by the national-level authorities referring to liquors of the best quality. The brands of the top ten liquors are mostly connected with their place of production and each brand has a story involved showing the depth and profoundness of the liquor culture of China.

1) Maotai

Maotai is named after the place where it's produced. The northern Guizhou[②] is known for its high-quality water and pleasant climate. The locals are good at making

① Yang Shuqiong. Chinese Character for "Wine" and "Wine Culture" [J]. *Journal of Inner Mongolia Radio & TV University*, Issue 6, 2017: 33 - 36.

② Northern Guizhou: the northern part of Guizhou.

liquor; hence the area is called the land of liquor. Among all the places in this area, the liquor made by Maotai Village of Huairen County is the best in taste which is called "Maotai burning" or "Maotai spring". Due to the superior quality of liquor produced in Maotai Village, it's widely known that the liquor produced in Maotai Village is hard to duplicate in other places. Therefore, whenever Chinese liquor is mentioned, people would say the best Chinese liquor comes from Maotai Village. As a result, this variety of Chinese liquor is normally called "Maotai liquor" or simply "Mao liquor".

Maotai liquor

2) Wuliangye

During the later years of Ming Dynasty and the early years of Qing Dynasty, liquor making industry had taken shape in Yibin. Deng Zijun, owner of "Lichuanyong" roast liquor workshop used red sorghum, long-grain rice, glutinous rice, wheat and maize as raw materials and made "mixed cereals liquor" which had rich aroma. In 1909, a large number of celebrities and scholars gathered for a banquet. When the bottle of the "mixed cereals liquor" was opened, the aroma was overwhelmingly intoxicating. Yang Huiquan, a juren of the late Qing Dynasty, suddenly asked, "What is the name of the liquor?" After hearing the description by the workshop owner, Yang Huiquan suggested, "For such a fine liquor, the name of 'mixed cereals liquor' seems to be in poor taste. As the liquor is made from five types of cereal, why not call it 'Wuliangye' (five cereals liquor)?" "Yes, it's indeed a good name." All those present expressed their admiration for the idea, hence the name of "Wuliangye" was formally created.

Wuliangye

3) Luzhou Laojiao

In 1996, the State Council designated the 400-plus years old cellar group of Luzhou Laojiao as the national key cultural relics protection unit. Built during the reign of Wanli Emperor of the Ming Dynasty (1573 AD), the cellar group is the earliest and best preserved in China which is still used today. Praised as the "living cultural relics", the cellar group is the precious heritage of the Chinese nation with inestimable cultural relics value, social value and production value. Luzhou Laojiao is the standard-grade liquor for Chinese liquor appreciation, hence the name "National Cellar 1573".

4) Yanghe

Yanghe liquor is named after its production place. Yanghe township in Suqian of Jiangsu Province was renowned for liquor making during the Han Dynasty. Originally called White Yanghe, there used to be a spring at the bend of the river which was named "Beauty Spring". Where was the source of the White Yanghe and how did Beauty Spring come into being? There are numerous tales having been told over the years. Made from the water from the Beauty Spring, Yanghe Daqu is clear, refreshing and sweet with long-lasting aftertaste. It tastes best when paired with boiled salted duck and Jinling roast duck of Nanjing.

5) Fenjiu

Fenjiu is produced at Xinghua village in Fenyang of Shanxi Province. There was a local liquor called "Fenqing" more than 1,400 years ago. Back then, there was no distilled liquor in China. "Fenqing" and "Dry Brewing" as recorded in historical documents are actually variants of millet liquor. Chinese liquors including high-grade liquors such as Fenjiu have all been derived from millet liquor. In the poetic masterpiece "Tomb-sweeping Festival", Du Mu, a poet of the late Tang Dynasty, writes, "Where can a wineshop be found to drown his sad hours? A cowherd points to a cot amid apricot flowers." Fenjiu hence became well-known. In 1915, Fenjiu won a gold medal at Panama World Expo winning honor for China and being recognized as the leading brand in China's liquor making industry.

6) Langjiu

Langjiu is produced in Erlang township in Gulin County of Sichuan. As a Fengshui treasure land with unique charm, the town is located at the middle reaches of the Chishui River and surrounded by high mountains. During the later years of the Qing Dynasty, the locals found a spring among the high mountains and deep valleys. Named "Lang Spring", the water from the spring is clear and sweet. The locals used water from "Lang Spring" to make liquor which was then called "Langjiu". Two natural cellars are hung near the Langjiu plant: Tianbao cellar and Dibao cellar. It's warm in winter and cool in summer in the cellars and temperature there remains at 19 degrees Celsius throughout the year. Langjiu stored in the cellars ages more quickly and is richer and aromatic. Among all the Chinese liquor manufacturers, Langjiu manufacturer is the only one to store liquor in natural karst caves.

7) Gujing Gongjiu

With distinctive style featuring "crystal-clear color, orchid fragrance, mellow taste and long-lasting aftertaste", Gujing Gongjiu has won four medals consecutively at national liquor competition. History of the liquor can be traced back to 196 AD when Cao Cao offered "Jiuyunchun liquor" (liquor produced in Bozhou, Cao Cao's hometown) and the production techniques to Liu Xie, Emperor Xian of Han. Emperor Xian of Han spoke highly of the liquor which was then designated as tribute liquor for the imperial house. During the Northern and Southern Dynasties, an ancient well was found in Bozhou, Anhui. Fragrance of the liquor made by using water from the well wafted far away and the well gained considerable fame. Since then, liquor-making workshops had sprung up in Bozhou and neighbouring areas.

8) Xifeng liquor

Produced in Liulin township in Fengxiang County of Shaanxi, Xifeng liquor was first produced during the Yin-Shang period and became popular during the Tang Dynasty and Song Dynasty. The production place was one of the settlement areas of the ancestors of Chinese people and the legendary birthplace of phoenix. In March during the Yifeng reign of the Tang Dynasty (676 – 679), Pei Xingjian, assistant minister of the Ministry of Personnel, escorted prince of Persia back home. When they arrived at Tingzitou village to the west of Fengxiang County, they witnessed a spectacular scene where bees and butterflies of Tingzitou village were intoxicated with aroma of the liquor stored in cellars of Liulin township, a place five miles away from Tingzitou village. Amazed, Pei Xingjian chanted a poem to express his admiration. Since then, Liulin liquor was designated as tribute liquor thanks to its "high quality, clear and rich texture and strong aroma". In modern times, Liulin liquor was renamed Xifeng liquor.

9) Dongjiu

Dongjiu is produced in Zunyi of Guizhou Province. During the Wei, Jin, Southern and Northern Dynasties period, liquor making industry was flourishing in this place with liquor infused with natural plants. As over 130 kinds of natural Chinese medicinal herbs are involved in the liquor production process, Dongjiu is a liquor of herbs and inheritor of the liquor production tradition of "homology of medicine and food" and "homology of liquor and medicine". The Chinese character for "dong" ("董") is composed of two radicals: "艹" and "重". "艹" means "grass" and "重" means a great quantity. The Chinese character of "董" means tradition, orthodox, uprightness, dignity and stateliness. The cultural connotations of the Chinese character of "董", cultural connotations of Dongjiu and the original production place of Dongjiu (Donggong Temple) have something in common. Therefore, the liquor was officially named "Dongjiu" in 1942.

10) Jian Nan Chun

Jian Nan Chun is produced in Mianzhu County, Sichuan. During the Tang Dynasty, Mianzhu County was known for its "Jian Nan Shao Chun" liquor. Legend has it that Li Bai, the great poet, once sold his fur-lined jacket in order to have enough money to buy the fine liquor and drink to his heart's content. The legend just reflects how enticing the liquor is. "Mianzhu Daqu" is the predecessor of Jian Nan Chun. In 1950s, Pang Shizhou, a poet based in central Sichuan, taught at Sichuan University. One day, he invited some cross-generation friends to his house to savor Mianzhu Daqu. One friend suggested changing the name of the liquor into Jian Nan Chun. Three days later, Mr. Pang gave three written Chinese characters "Jian Nan Chun" to Mianzhu Distiller. He also offered interpretation of these three characters:

The two characters for "Jian Nan" suggest that the liquor is produced in Mianzhu which is located to the south of Jianmen Xiongguan. People would associate the name with the fertile Tianfu plain. The character for "Chun" is an adaptation of ancient form for present-day use. Su Dongpo once said that liquors in the Tang Dynasty mostly used "Chun" for brand name. The character allows people to enjoy the fine liquor and connect it with the delight of the spring.①

5. Wine and calligraphy

As shown by the Chinese character of "jiu" written in the style of oracle-bone script or jinwen and the exquisite calligraphy printed on the packaging box of famous liquor such as Maotai and Wuliangye, wine culture in China is tightly tied to the art of calligraphy. The origin, calligraphy and contents of calligraphy work of "Lanting Xu" are all closely related to wine. Many famous calligraphers over the centuries wrote calligraphy due to their wine addiction creating splendid scenery by fusing the color of ink and the aroma of wine. Their works have great artistic and historical value because they expressed their real sentiments through their works and these works are not reproducible.②

Most calligraphers and painters in China had wine addiction. Two reasons behind this phenomenon: (1) These artists were romantic and melancholy. With strong personality, they usually drank wine in order to realize their dreams. (2) Wine can stimulate central nervous system making people excited and spark inspiration for artistic creation. This is why the legend of "Libai composed 100 poems after drinking 10 liters of wine" is still told today.

"Lanting Xu", the finest calligraphy work by Wang Xizhi (303 - 361 AD), "Sage of Calligraphy" during the Eastern Jin Dynasty, was written when he was

---

① Li Lin. *Introduction to Tea and Wine Culture*[M]. Taiyuan: Beiyue Literature and Art Publishing House, 2010: 164 - 175.

② Hou Zhongming. Chinese Calligraphy and Wine Culture[J]. *Journal of Sichuan University of Arts and Science*, Issue 1, 1999: 64 - 66.

drunk. As the most notable work in China's calligraphy history, "Lanting Xu" is reputed to be the "Best Calligraphy Work in Running Script". Motive for writing "Lanting Xu" originated from an outdoor wine game: Wang Xizhi and his friends (all of whom were poets) gathered at Lanting in Zhejiang. They sat beside a stream dug from a river. Cup filled with wine was placed at the upper reaches of the stream. The cup then flowed down the stream. When the cup stopped in front of someone, he must compose a poem on the spot; otherwise, he had to drink the wine as a punishment. In total, they composed 37 poems and decided to arrange the poems into a collection. They selected the venerable Wang Xizhi to write the preface to the collection. Drunk and with all kinds of feelings in his heart, Wang Xizhi wrote a 324-character preface expressing his joy and sadness. The finished calligraphy work is "Lanting Xu", a celebrated work in running script which has a deep impact on Chinese calligraphy for more than 1,000 years. After Wang Xizhi sobered up, he tried to copy the original work 10 times but none of them can surpass the original which he had written when he was drunk.[①]

Huai-su, a monk in the Tang Dynasty, loved drinking. After getting drunk, he would write calligraphy on temple walls, screens, clothes and utensils and therefore was called "drunk monk". "Autobiography", his calligraphy masterpiece, was written when he was drunk. Li Bai once described Huai-su as: "My teacher leaned beside bed after getting drunk and wrote several thousand pieces of calligraphy in a moment. Ink is spattered on paper like a swift rain and paper falls on the ground like dancing snowflake."

Zhang Xu, a master calligrapher of the Tang Dynasty, is known for his "wild cursive script". After getting drunk, he was overcome with inspiration and excitement. Shouting wildly and walking quickly in the courtyard, he used a brush or even steeped his hair with ink to write calligraphy. After sobering up, he himself

---

① Hou Zhongming. Chinese Calligraphy and Wine Culture[J]. *Journal of Daxian Teachers College*, Issue 1, 1999: 64-66.

even felt amazed at his calligraphy, hence the creation of the calligraphy work of "Four Calligraphy Works on Poems" whose characters "looked like cloud and mist spread on paper.①"②

Stories of these master calligraphers are not intended to overstate the role of wine in the art of calligraphy. The masterpieces of these calligraphers were created on the basis of their mastery of traditional calligraphy skills. Wine addition alone can't lead to creation of calligraphy masterpieces.

6. Wine and martial arts

Throughout history, only young men are expected to perform military service. Loneliness in the military barracks, cruelty of war, unpredictability of fate and threat of death create psychological needs for wine. Wine can lessen mental stress, relieve fear and comfort or intoxicate restless soul. The invigorating effects of wine can also enhance morale of the troops before battle. The defense and offense skills learned from the wars and the improvement of defense skills constitute the foundation of Chinese martial arts.

It's said that poets are usually wine lovers. In fact, martial arts practitioners also love drinking because wine drinking can demonstrate their forthrightness and martial arts spirit and at the same time, express their aspirations. More importantly, wine is the "magical liquid" to help them achieve excellent martial arts skills. Fu Qingzhu, the famous founder of Fujiaquan of the Qing Dynasty, created drunken boxing techniques when he was drunk.

Fu Shan, style name Qingzhu and alias Qiao Huang, was born in 1607 and died in 1684. He was a famous thinker, poet, scholar, painter and patriot during the later years of the Ming Dynasty and the early years of the Qing Dynasty. He was

---

① A line from the poem of "Eight Immortals Drinking Wine" by Du Fu which praises calligraphy of Zhang Xu, a metaphor meaning Zhang Xu's calligraphy is like cloud and mist spread on paper.

② Liu Yong. *Chinese Wine (Chinese and English Version)* [M]. Hefei: Huangshan Publishing House, 2012: 147 - 149.

also skilled in martial arts. His boxing techniques have become a school of martial arts enriching the treasure house of Chinese martial arts. He integrated martial arts with painting. It's said that he usually painted pictures after getting drunk. Being alone in a room, he would practice martial arts for a while before painting pictures. By practicing boxing while getting drunk, Fu Qingshan became unconscious of the boundary between himself and the external world. He then incorporated his feelings into his paintings. As a result, his paintings are solemn in style as if a storm is approaching and at the same time, his paintings are lively and transcendental in spirit. His boxing techniques also look like the movement of a drunken person. Fu Qingshan finally created the fusion between wine, painting and martial arts.

"Drunken boxing" is an important style of modern performance martial arts. Its name derived from its skills and gait which look like the movement of a drunken person. The drunken style and postures were borrowed from the ancient "drunken dance". Its attack tactics incorporate those of other styles of martial arts. With both strength and grace, drunken boxing features feigned attacks and various techniques. It became an established taolu during the Ming and Qing Dynasties. It's practiced in various regions in China including Sichuan, Shaanxi, Shandong, Hebei, Beijing, Shanghai and areas around Huai and Yangtze Rivers.

Drunken boxing practiced today can be classified into two groups: ancient drunken boxing taolu which emphasizes effectiveness of attacks; modern drunken boxing taolu which highlights drunken movements including falling, pouncing and rolling. Both groups must adhere to the rule that the practitioner seems drunk but in fact remains somber in heart. The drunkenness is just an imitation rather than a real drunken state. During fight, the practitioner staggers and seems unable to stand firmly because of drunkenness. However, while staggering or rolling, he or she would launch attacks unexpectedly when there is opportunity. This is the key to power and effectiveness of drunken boxing.

It seems that drunken boxing does not follow any rules. Actually, it's not true. How can a drunken man practice martial arts if he does not follow any martial arts rules? Every movement of drunken boxing strictly complies with rules emphasizing the changes of and coordination among hands, feet, footwork and direction of eyes. The graceful and unrestrained movements of drunken boxing practitioner are the results of diligent training which are then presented with proficiency.

In addition to drunken boxing, there is another style of martial arts called "drunken sword". Rich in cultural connotations, swordsmanship has a long history in China. As an ancient weapon made during the Neolithic period, sword is still used today for martial arts practices. It once served as symbol of imperial authority, instrument for the immortals to cultivate energy and tool for the intellectuals to express their feelings and aspirations. Today, it's also used by artist as stage property to portray dramatic character.

In terms of taolu, there are over 30 types of swordsmanship including Taijijian, Taiyijian, Baxianjian, Baguajian and Qixingjian. Different kinds of swords are used for martial arts practice including single sword, double swords, sword with long tassel, sword with short tassel, sword wielded single-handedly, sword wielded with both hands, sword used individually and sword used for one-on-one practice.

One style of swordsmanship is drunken sword which is based on wine culture. Due to its unique style, drunken sword has gained widespread popularity. Drunken sword is often practiced for performance purpose and integrated with drama performance and dance performance. Drunken sword has the following characteristics: unrestrained style; quick movement intertwined with slow movement; adopting unexpected tactics. It seems that the movement does not follow any rules as if the practitioner is drunk. Actually, each staggering and rolling movement hides a forceful attack.

As a weapon, sword had disappeared from battlefield. It's primarily used as physical exercise apparatus. Swordsmanship is now a performance combined with dance. As drunken sword imitates the movement of a drunkard and features unexpected tactics and dynamic movements, it occupies a unique place in the field of swordsmanship.[①]

① Li Lin. *Introduction to Tea and Wine Culture*[M]. Taiyuan: Beiyue Literature and Art Publishing House, 2010: 234 - 242.

Chapter 6 Banquet Culture

Banquet is a formal dining event for socializing purposes such as welcoming someone, expressing appreciation, offering congratulations and celebration. Banquet links people in accordance with etiquette rules. It's an enjoyment in everyday life. It's also one of the important indicators of the progress a country has made in material well-being and cultural & ethical advancement. Modern Chinese-style banquet has three basic characteristics:

(1) Wine is the centerpiece of a banquet. As the saying goes, "no wine, no banquet." Holding a banquet in China is called "offering wine" and attending a banquet is called "drinking wine". Wine can stimulate appetite and enhancing joyous atmosphere. Therefore, menu of a banquet is usually arranged by following the rules of "wine is the soul of a banquet" and "dishes are arranged for wine drinking".

(2) Emphasizing seating arrangement and dining etiquette. "Holding a banquet for guest of honor; etiquette is essential for a banquet." Etiquette is an important element of banquet culture in China. Seating is carefully arranged for a formal banquet and no one is expected to sit wherever he likes. Seating is usually arranged according to seniority in the family, social status and closeness of the guest's relationship with the host.

(3) Varieties of dishes served in prearranged order. Dishes are various including hot dish, cold dish, meat, vegetables, salty flavor, sweet flavor, thick taste, light taste, crispy food, soft food and hard food. Dishes are served in predetermined order. A typical Chinese banquet consists of three phases. Phase one is the "overture" which comprises

many dishes such as snacks, appetizers, first course of soup and cold dishes. Phase two is the "theme" where the main courses and hot dishes are served including main course, roast (fried) dishes, dishes made with reused soup stock, fish, meat, vegetables, sweet food and individual soup. Phase three is the "epilogue" when rice or noodles and seasonal fruits are served.

The design of modern Chinese-style banquet follows the principles of "proper selection of flavors and appearance" and "proper selection of raw materials" in order to highlight the theme of the banquet, the creativity of dishes and the interesting aspects of names of courses.

All in all, Chinese banquet culture has a long history, rich connotations and high artistic value. Banquet art is the most perfect embodiment of the culinary art of China and one major component of world arts and cultures. To elaborate on the banquet culture of China, this chapter will focus on dining etiquette, banquet dishware, wedding feast, hundred-day feast, old person birthday feast, white feast, family feast, graduation feast and the six banquets of Manchu Han Imperial Feast.

6.1 Dinning Etiquette

Dining etiquette is an important element of modern Chinese-style banquet. It consists of the following aspects:

1. Seating arrangement

Chinese-style banquet mostly uses round table and seating is arranged according to seniority and superiority of guests. Host of the banquet sits at the main table facing the entrance. As for the seating arrangement for the same table, those who are superior sit nearer to the host and those are inferior sit farther away from the host. For those who sit with the same distance to the host, those sitting on the right hand of the host is superior and those sitting on the left hand of the host is inferior. If a banquet uses several tables, each table must have a host who is the representative of the host of the main table. The representative host sits facing the same direction as the host of the main table or sits facing the host of the main table. A maximum of 10 guests will sit at one table and the actual number of guests shall be an even number. It'll be crowded if too many people sit at one table and the server will get too busy.

2. Table manners

Whatever the flavor, standard Chinese-style banquet follows the same course serving order: cold dishes; stir-fried dishes; main course; dim sum; soup; fruit platter. When only one third of food is left in cold dishes, the first course is served. No matter how many tables are used for a banquet, the same dish or course shall be served for all the tables at the same time.

If dishes are served by the server, the server shall serve the honor of guest first and then serve the host and at the same time, women shall be served first. The server may also serve clockwise. If the guests serve themselves, each course shall be placed in front of the guest of honor and all the guests, starting from the guest of honor, serve themselves clockwise; otherwise, it'll be considered impolite.

Before a banquet starts, the server will offer each guest a wet towel to wipe hand. By the end of the banquet, another wet tower will be offered to guests to wipe mouth. The server also prepares a napkin for each guest. The napkin shall be spread on the lap. It shall not be worn around the neck or tucked into collar or belt. The napkin can be used to wipe mouth and hand gently, but it can't be used to wipe dinnerware or wipe away sweat.

Due to the characteristics and culinary traditions of Chinese-style banquet, some rules shall be followed when attending a banquet: (1) Do not lift your chopsticks when dishes are served. Wait until the invitation from the host and lift your chopsticks only after the guest of honor does so. (2) Be courteous to each other when taking food from the dishes; taking food in turn and avoid fighting.

(3) Taking appropriate amount of food; do not eat up your favorite food alone. Asking others to enjoy the food first so as to show hospitality; do not grip food or refill bowl for others without considering their preferences so as to avoid embarrassment. (4) Avoid being picky about food; do not pick and choose food from the communal plate continuously. (5) Taking food from the plate immediately after gripping the food; do not drop the food onto the plate after gripping it or returning the food to the plate after taking it.

3. Rules on dining utensils

Chopsticks are the most widely used utensils for Chinese-style banquet. Use right hand to hold chopsticks and it's incorrect if chopsticks are placed too high or too low between fingers. The correct way of using chopsticks is placing the thumb at 1/3 point of the chopsticks (measuring from the tip of chopsticks), which looks graceful and makes it possible to open and close chopsticks. Rapping an empty bowl with chopsticks is a taboo behavior in China because beggars would rap bowl in this way. It's also considered a taboo if sticking chopsticks into a bowl of rice. This is a practice during a funeral when people worship ghost; therefore, the behavior is considered ominous as it would remind people of the deceased.

6.2 Dinnerware and Flatware for Banquet

Dinnerware and flatware used for a modern Chinese-style banquet includes bowl, plate, saucer, spoon, chopsticks, chopsticks holder and stewed food pot. Normally, the size of a utensil shall be suitable for the amount of food, type of utensil shall match that of the food, color of utensil shall match that of the food and utensils shall match with each other.

Over the several thousand years of history of China, china dinnerware has been widely used for nearly 2,000 years. As a significant contribution to world civilization, "china" also refers to the country of China. Features of china include: hard; white porcelain body; fine texture; lightweight and unable to absorb water; exquisite appearance; clean; durable; high quality and inexpensive; easy to store.

Bowl is one of the widely used dining utensils. Bow and chopsticks constitute the basic dining method for people of the East Asia. Plate, pot and small plate are also widely used dining utensils in China. Similar to bowl, plate, pot and small plate used today are presented in various appearances. Round, square and rectangular utensils are found at many occasions. Using different shapes of beautiful dining utensils for modern banquet can enhance the atmosphere and stimulate appetite.

Cup and pot are used to contain drink. You can drink directly from a cup. For example, wine cup is used for drinking alcoholic drink while tea cup is used for enjoying tea. Pot can be used as a medium to pour tea or wine into cup, or can be used by an individual as a drinking utensil. At Chinese-style banquet, pot is mostly used as a medium and cup is used as drinking utensil by an individual.

Over the centuries of cultural evolution in China, chopsticks, spoon, table-knife, fork and soup spoon had all been used as tools for gripping food. Knife, fork and spoon had been used for dining purpose during the culinary history of China. Finally, hand, chopsticks and soup spoon have become the most common way to take food at table.

Invented by the ancient Chinese, chopsticks are unique in the world. As a symbol of the long history of Chinese civilization, chopsticks are the quintessential component of culinary culture of China. Since ancient times, chopsticks have been made of gold, silver, copper, iron and stainless steel. Most chopsticks used for Chinese-style banquet are made of wood or bamboo.

Standard way of holding chopsticks shall be followed. It's incorrect to hold chopstick too high or too low, or use different fingering. When using chopsticks to pick up food, one should avoid picking and choosing food continuously or using chopsticks to pierce food. If other people also pick up food at the same time, behave

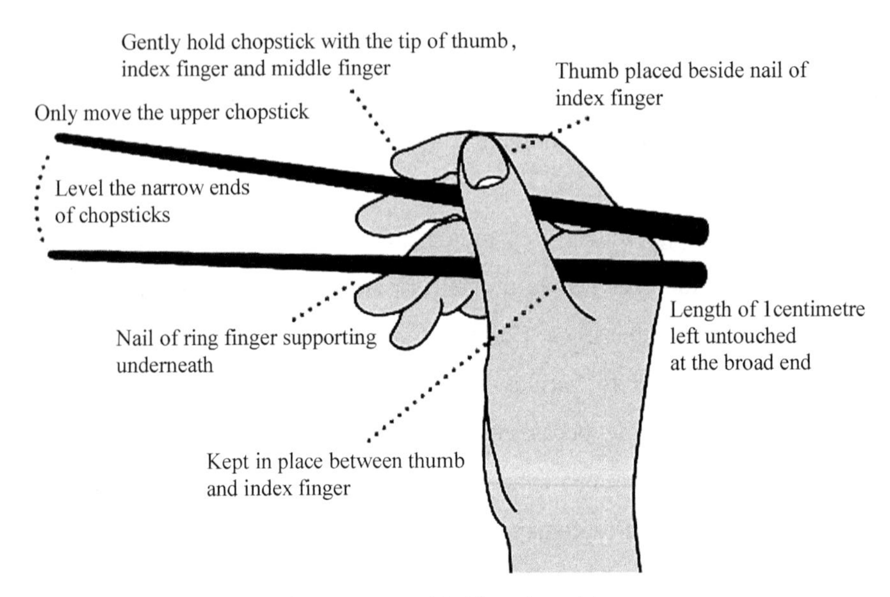
Gently hold chopstick with the tip of thumb, index finger and middle finger
Thumb placed beside nail of index finger
Only move the upper chopstick
Level the narrow ends of chopsticks
Length of 1centimetre left untouched at the broad end
Nail of ring finger supporting underneath
Kept in place between thumb and index finger

Standard way of holding chopsticks

courteously and pay attention that your chopsticks do not touch chopsticks of others. It's an ungainly sight that if you put chopsticks in your mouth or use chopsticks as toothpick. When engaging in a conversation, it's considered impolite if you use chopsticks as a tool and point at the others. At the end of banquet, chopsticks should be put back in a neat manner and leave the table after all the guests put down chopsticks.

6.3 Wedding Feast

Wedding feast, also called "drinking at wedding feast", is the occasion when bride and bridegroom invite relatives and friends to gather to celebrate their marriage and express their gratitude to the guests. Today, most wedding feasts are held at restaurants. In a few places, wedding feasts are held in the courtyard of the family concerned or in open space. Before the wedding feast begins, the bride and bridegroom would stand at the entrance to the banquet hall welcoming the guests and expressing their gratitude. The feast formally starts when all the guests have arrived. An emcee acts as the host for the wedding ceremony. The emcee would invite the bride and bridegroom onto the stage and tell the guests about the love story of the bride and bridegroom. Sometimes, their parents would be invited onto the stage as well expressing their best wishes to the couple. As for seating arrangement, family members, relatives, former classmates and colleagues of the bride or the bridegroom sit at one table. Or those who are of the same age and who are familiar with each other sit at one table since they may have a great conversation and thereby increasing the joyous atmosphere of the wedding feast. Those sitting at the main table are guest of honor (boss of the bride or bridegroom), parents of the bride and bridegroom, the bride and bridegroom, bridesmaid and groomsman.

The banquet begins after the end of the wedding ceremony. The couple and all the guests start to enjoy the food. Then, the couple stands up and, accompanied by the bridesmaid and groomsman, they drink a toast to guests of each table. By the end of the feast, the bride and bridegroom stand again at the entrance to the banquet hall shaking hands with departing guests and saying goodbye and other words such as "thank you for coming" and "take care".

Holding a wedding feast is a common practice in most places in the world. However, procedures and activities of the feast vary depending on local customs. As China is a country with vast territory and many ethnic groups, customs of different provinces, cities and autonomous regions are also different.

Features of menu of a wedding feast include:

(1) Even number of dishes. Most places in China offer an even number of dishes at "red happy event (wedding feast)" and an odd number of dishes at "white happy event (funeral feast)". Eight dishes on the menu symbolize making a fortune, ten dishes symbolize perfection and twelve dishes symbolize happiness for each month.

(2) Name of dishes is usually auspicious word expressing good wishes to the couple. Examples include: "staying together like fish and water" (fish ball with

milk soup); "having a baby" (red dates, peanuts, longans and lotus-seed soup); "flying side by side" (double pearl shrimps). A chicken dish symbolizes luck and happiness. A fish dish symbolizes abundance and prosperity year by year and is usually served as the last course. Fruits usually include pomegranate (more sons, more blessing; booming) and peach (sweetness and happiness). However, pear does not appear on the table as the Chinese character for "pear" sounds like the Chinese character for "separation" ("li").

(3) Habits and dining taboos of guests are taken into consideration when arranging wedding feast. National customs, religious belief, preference and taboos of guests should be fully taken into account. For examples, guest believing in Islam does not eat pork.

The following is a reference for menu arrangement for a wedding feast:

One assorted dish: happy couple (patterned cold dish).

Four surrounding dishes: heavenly maids scattering blossoms (fruits and flowers carving); matchmaker offering fruits (dried fruits and sugar-preserved fruits); three stars shining on high (assorted meat); good luck arrive in succession (assorted vegetables).

Ten hot dishes: happily married couple (pipa duck flippers); kylin sending son (kylin mandarin fish); fated marriage (three shreds egg roll); a perfect pair (shrimp balls and green beans); happy union (ham and melon soup); perfect son-in-law (braised pork with bamboo shoots); flying side by side (spiced crisp quail); nuptial knot (skewered mutton); beauty washing silk (boiled cabbage); handsome man cultivating (sweet corn soup).

One soup dish: oath of marriage (hotchpotch).

Wedding feast is an important part of wedding ceremony. Due to differences in regional habits and customs, wedding feast procedures in China have become diversified. This section introduces modern wedding feast in China. As traditional wedding feast is more complicated in procedures, this section will not look at it in detail.

Wedding feast

6.4 Hundred-day Feast

Hundred-day feast is a celebration held 100 days after a baby was born. As an ancient custom in China, hundred-day feast is still popular across the country. The host invites guests to a banquet. Guests present gifts and the host present gifts in return to share love and joy. The purpose of "hundred-day feast" is to wish the baby happiness and longevity. In ancient times, "hundred-day feast" was also called "many blessings". As the Chinese character for "blessing" sounds like the Chinese character for "deer", the traditional way of expressing best wishes is to present a picture depicting patterns on deer as a gift. Titled "hundred patterns and many blessings picture", the picture symbolizes best wishes for the baby.

Rules on hundred-day feast vary. There used to be the custom of "wearing clothes from 100 households" and "eating food from 100 households". Similar to giving a humble name to a baby, the above-mentioned custom was intended to borrow blessings from the others. Therefore, family with a very old person is usually expected to "lend longevity", i.e., lend a dress or a handful of rice to the baby to increase the baby's longevity. Today, the custom is still adopted in cities. Though only the formality is followed, the original intention is somewhat kept. Due to increasingly higher standards of living, babies are usually physically strong. However, the custom is still followed out of people's fear of fate and their wishes for happiness for baby.

Atmosphere of hundred-day feast is usually lively, lovely, warm and vivid. If the baby is a girl, pink color is usually the predominant color such as pink, light yellow, light purple and white. If the baby is a boy, solid color is the predominant color such as blue, white and yellow.

Dishes of a typical hundred-day feast include: cold dishes; hot dishes; soup; drinks; dim sum. Cold dishes include: botou chicken cooked in wine, yellow croaker in wine sauce; braised duck with soy sauce in Shanghai style; marinated pork tripe; cold evergreen dressed with sauce; fish jelly; crispy cucumber; steamed bean curd roll. Hot dishes include: steamed scallop with fermented soya beans; steamed sturgeon; scalded prawns; pork knuckle; boiled eel back; bullfrog on iron plate; stir-fried beef fillet with tea tree mushrooms; eggplants with garlic sauce; stir-fried crab with egg yolk; "West Lake" thick beef soup; golden garlic chicken; rice steamed pork ribs; flounder fried with potherb mustard; pumpkin, lily bulb and yuan xiao soup; ruyi 100-day noodles. Soup includes: fairy duck pot; goose pot. Dim sum include: fried noodles; peach-shaped birthday cake; cream cake. Drink includes: orange juice; red wine; brewed wine.

Dishes offered at hundred-day feast are in even number symbolizing good luck and harmony. A minimum of eight courses are served for a table of ten guests. Normally, twelve courses are more acceptable since "twelve" symbolizes happiness for each month. Sometimes, sixteen courses are served if there is small amount of food for each course. The principle of "no waste" is followed when deciding the number of courses served.

Deliciousness, adequate amount, sumptuousness and freshness are judging criteria for the quality of food served for hundred-day feast. Dishes include meat dishes, vegetables dishes, fish, meat, cold dishes and hot dishes. If seafood is offered, the feast will look upscale and sumptuous. If all these standards are met, guests will be full of praise.

As for the arrangement of menu, it's advisable to consider child's taste as it's a feast held for baby. To show the loveliness of child, the menu can include such items as steamed stuffed buns of various lovely shapes, Mickey-shaped rice or little bear cakes.

As for the taboos of hundred-day feast, suckling pig dishes should not be served. According to popular belief, suckling pig is a baby pig and therefore it's inappropriate to eat suckling pig at a hundred-day feast as it bodes ill. Eating fish having fish eggs is considered cruel as the mother gives birth to the baby after 10 months' pregnancy. In some places, chicken should not appear at hundred-day feast if the baby is a rooster in Chinese zodiac. As China is a country of vast territory and many ethnic groups, customs vary significantly across the country. Therefore, it's advisable to make menu arrangement on the basis of local customs.

Compared with wedding feast, only a small number of families hold hundred-day feast and a considered number of families do not hold hundred-day feast. Baby is too young to be in the middle of the adults or a large group of people considering the baby's hypoimmunity. Therefore, some rules must be followed when holding or attending a hundred-day feast.

(1) If the baby is also present at the feast, guests should not cuddle or kiss the baby; otherwise, baby will feel insecure and is likely to be infected with germs or disease. To ensure baby's physical and mental health, it's advisable to reject request of guests politely if they want to cuddle baby.

(2) Do not feed baby at the feast. Baby less than one year old can't eat food containing sugar or salt. As baby has not started to grow teeth, they can't chew most

Hundred-day feast

food items of the feast, especially nuts which may cause chocking and is therefore life threatening. To distract baby's attention from the food on the table, parents may bring food and toys from home.

(3) Air-conditioner is usually turned on in the banquet hall creating low temperature in summer and high temperature in winter. To ensure baby does not catch a cold, put on clothes or take off clothes for baby when going into or leaving banquet hall.

(4) Considering the limited energy of baby, the baby must take a rest immediately after the feast. Baby will get exhausted if the feast lasts too long. Do not interrupt baby's daily schedule.

6.5 Old Person Birthday Feast

Old person birthday feast is an occasion when guests offer birthday congratulations to an old person. Regions vary in customs but a feast is normally held for the occasion.

In ancient time, people didn't celebrate birthday. According to the "filial piety" tenet of Confucianism, "parents have suffered a lot to bring me up." Therefore, it's supposed to reflect on the hardships the parents had endured in order to bring me up. That's why no one celebrated birthday or offer birthday congratulations in ancient times. The practice of celebrating birthday was adopted during the Southern and Northern Dynasties. *Family Instructions of Yan Clan* contains records on holding banquets to celebrate birthday every year. Interestingly, both the habits of celebration of birthday and no celebration of birthday stem from the concept of "filial piety": no celebration of birthday is to reflect on the hardships parents have endured; celebration of birthday is to make parents happy. During the Tang Dynasty, birthday celebration was a joyous event when banquet was held, music was played and birthday congratulations were offered. Since then, birthday celebration custom with offering birthday congratulations as the purpose and banquet and entertainment as the form of celebration has been practiced until today. Since the Song Dynasty, the practice of presenting gifts on one's birthday became popular and is still adopted today, constituting an important part of birthday celebration.

An old person was called "longevity" in ancient times and "longevity" stands for a long life expectancy. To fulfill filial duties, children would hold grand celebration activities on their parent's birthday. Relatives and friends were invited to attend the birthday feast and the guests would offer their birthday congratulations. Priority given to old person birthday celebration is an embodiment of the traditional virtue of the Chinese nation which emphasizes respect for the elders.

Today, 50 or 60 years old is a dividing line and only those who are over 50 or 60 are qualified to hold an old person birthday party. In most places, only those who are over 60 can hold an old person birthday party, hence the saying, "celebration the 60th birthday". In some places, those who are 40 years old may hold an old person birthday party.

Preparation for celebration starts on the eve of the birthday. On this day, the daughter and daughter-in-law return to the wife's parents' home bringing gifts. They would have dinner with the one whose birthday is celebrated and other family members. To prepare for the next day's celebration, birthday celebration hall is arranged and gifts from the children as well as relatives and friends are displayed in the hall. On the day of the parent's birthday, relatives and friends are invited to attend the banquet and the ancient ritual of kowtow is usually performed. The ritual of kowtow is popular in the countryside, especially in remote rural areas. In cities, celebration feast is mostly held in restaurant or hotel. Birthday celebration hall is arranged at the site of celebration and celebration ceremony and feast are held concurrently. Ceremony may consist of different number of procedures. Style of ceremony can be a combination of the ancient and the modern, or a combination of the traditional Chinese style or the Western style.

Birthday feast is an important element of celebration. The host holds a banquet to entertain guests. Items on the menu include fine dishes and costly food, and "birthday noodles" is a must-have. All the guests bring gifts such as peach-shaped birthday cake, birthday cake, birthday noodles, birthday candles, birthday couplets, birthday cash gift and longevity umbrella. If possible, these gifts are usually decorated with patterns and pictures symbolizing longevity.

Eggs, snacks and birthday noodles are normally served first. The old person celebrating birthday does not sit at the table in the main hall. Instead, he or she would sit at another table in the inner room accompanied by several old people of similar age. Dishes arrangement follows the principle of "the more, the better" symbolizing more blessings and longevity. Name of dish usually contains the Chinese character for "nine" and "eight" such as "double nine birthday feast" and "eight immortals dish". In addition to birthday noodles, peach-shaped birthday cake and birthday cake, gingko, pine nuts and red date soup are also offered at the feast. Name of dish is carefully crafted such as "eight immortals crossing the sea", "three stars gathering", "fortune as boundless as the East Sea" and "cloud and pine". Fish seldom appears on the menu. Watermelon soup, white gourd soup and stir-fried kidney are never served at the table. In some places at the lower reaches of the Yangtze River, the married daughter is supposed to offer birthday congratulations to her father or mother on their 66th birthday. She would cut pig ham into 66 bean-shaped chunks which are called "bean pork". After being braised in soy sauce, the pork chunks are then placed on top of a bowl of rice which is then placed in a basket along with a pair of chopsticks. Covered with a red cloth, the basket is then presented to the parent who would eat the pork, an act of birthday celebration. The custom of cutting pork into many chunks symbolizes longevity.

Banquet hall usually looks splendid with beautiful decoration. Table is covered with red tablecloth with lush green plants in the middle. Banquet hall is all decorated with lights and festoons with red lanterns as the adornment creating a joyous and vivid atmosphere in the room. On the wall, there are red paper with the Chinese character for "longevity" which are written by the children or grandchildren, or large birthday congratulations painting depicting pines, cypresses and cranes. These calligraphy works and paintings highlight the theme of the setting creating a beautiful and lively atmosphere. Sometimes, birthday couplets are pasted on the wall of the banquet hall. With a pair of poetry lines of "Sky has another year of existence and man has another year of life; spring comes and blessings are bestowed on the family", the couplets surround a big red Chinese character for "longevity". Under the big Chinese character, there are two chairs. On each side of the chair, there are two sacrificial tables (square table) of different sizes. On the table, there are birthday congratulations objects such as the three stars of Fu, Lu and Shou, fresh peaches. On the floor of the banquet hall, there are four big balloons containing numerous small balloons and a big gift pack tied with red ribbon. The gift pack is used for storing gifts presented by children and grandchildren. The background music is usually songs such as "A Better Future" and "Song of Birthday Congratulations". The old person celebrating birthday or his (her) children or grandchildren would call at the house of the guests who are the seniors in terms of age and status and express thanks, a custom called "a returned visit".

Whether it's hundred-day birthday celebration, or celebration of the birthday of a youth or an adult, or celebration of the birthday of an old person, the scale of celebration varies depending on the family's pecuniary conditions. Procedures of celebration are also different in different regions.

Old person birthday celebration is also popular among ethnic minority groups whose celebration is sometimes unique in some aspects. For example, when the Zhuang people hold old person birthday celebration, the younger generation would use pork and chicken to worship the ancestors. After the worship, they would cluster round the old person to sing the "Song of Birthday Congratulations".

Auspicious objects also appear at the occasion of old person birthday celebration such as peach-shaped birthday cake, crane, tortoise, pine, Five Auspicious Objects Painting, auspicious objects bearing the Chinese character for "longevity", birthday noodles, Chinese rose, gourd, longevity stone, the Southern Mountain, cat, butterfly and deer.

At the old person birthday feast, whether western style birthday cake introduced from the western countries, or the traditional Chinese food items such as birthday noodles, peach-shaped birthday cake or Chinese-style birthday cake, all of them are used to express best wishes for health, longevity and happiness.

The following is a reference for menu arrangement for an old person birthday feast:

Feast of "fortune as boundless as the East Sea"

Lucky sky shines on high (assorted dish)

Auspicious hall (poached shrimps)

Prosperity and longevity (crispy milk seafood rolls)

Golden rooster offer birthday congratulations (roasted crispy chicken)

Green old age (baked crabs)

Fortune as boundless as the East Sea (steamed garoupa)

Garden overflowing with beauty of spring (braised mustard with mushrooms)

Longevity (longevity bun)

Longevity and wealth (deep-fried noodles with goose liver paste)

Good luck year after year (pawpaw stewed with two snow pears)

Global fresh fruit (fruit platter)

Old person birthday feast

6.6 White Feast

White feast is an occasion for family members to remember the person who just passed away. Relatives and friends of the deceased are usually invited to attend the feast. "White affairs" is a term used to refer to handling funeral related affairs. In China, funeral affairs include attending upon a dying person, giving an obituary notice, encoffin, guarding coffin, keeping coffin, bereavement period, offering condolence, ceremony on the third day of the funeral, holding a funeral procession, burial and bereavement period. The funeral is a way of expressing good wishes for the deceased. In Chinese culture, the term of "red and white happy events" refer to wedding and funeral. White feast is an important part of the "white affairs".

When attending white feast, do not talk loudly and switch off cellphone (and other similar devices) or place it in vibration mode. Avoid exciting topics when chatting with relatives or friends at the feast. When family members of the deceased come to drink a toast, respond politely and say "my condolences".

If the deceased had enjoyed high prestige and had a flourishing family, he or she then had all the blessings. If the deceased was over the age of 60, he or she had enjoyed complete longevity. If the deceased died of natural causes rather than illness, he or she then died a natural death. If the deceased had all the blessings, had enjoyed complete longevity and died a natural death, his or her family hold a grand banquet to entertain relatives, friends and neighbors as a way of celebration. Through the happy event, people hope that their grief may be lessened and they themselves may continue their happy life. Therefore, atmosphere of the grand feast is usually lively.

The host also presents towel and bowl to guests as gift. As washing fabric, towel symbolizes getting rid of bad luck. It can also be used by relatives and friends to wipe off sweat and tears as some of them may help with funeral service matters. As article of everyday use, towel is inexpensive. For family holding funeral and white feast, buying towel for guests may help reduce expenses. If attending a funeral for an elderly person who died naturally or a funeral for a centenarian, guests are usually given a bowl which, according to popular belief, may bring "good luck of longevity" to them.

Dishes on menus should be in odd number rather than in even number. Nine objects are used at white feast symbolizing grief over loss of a family member. In ancient times, meat dishes such as fish were not supposed to appear on menu of white feast. Instead, vegetable dishes were served to remember the dead. Mutton is a must-have on the menu as sheep kneels down to suck breast, an act of showing gratitude to nursing. Therefore, eating mutton is a way of showing filial piety. In ancient times, tofu meal consisting of bean products and vegetables was prepared to remember the dead, and to entertain relatives and friends who had helped with funeral affairs with their money, labor and materials resources. To the ancient people, tofu was longevity food and Chinese works for "tofu" sound like the Chinese term meaning "many blessings". Serving tofu was then a way of expressing wishes for blessings and longevity for family members and guests. Today, in addition to tofu and vegetables, fine food and wine are also offered.

As China is a country with many ethnic groups, customs vary among different regions. To truly understand white feast culture of a place, the best way is conducting field investigation. For example, three animals (pork, fish and chicken) are first served at a white feast in Guangdong and the feast does not require rigid formalities. The last course of a Hakka feast is Chinese cabbage. As for wine, unlike wedding when any alcoholic drink can be served, white feast only serves millet wine and Chinese liquor. When drinking wine at white feast, avoid playing the finger game, talking loudly or engaging in drunken brawls. In Jiangsu and Zhejiang, those attending white feast can give white envelope to the host as a way of offering condolence. The amount can't be in even number or whole number. An additional one yuan can be added into the envelope.

6.7 Family Feast

Held at home, family feast is confined to home and is small scale gathering featuring diverse food items. Compared with formal banquet, family feast aims to create friendly and warm atmosphere allowing both host and guests to engage in casual conversation and thereby increasing mutual trust and understanding. No specific formalities are required for a family feast. To show respect and friendliness for guests, the hostess usually cooks food while the host acts as a server or vice versa. Both ways of entertaining guests may make guests feel at home.

Rules on family feast arrangement are as follows:

(1) Taking season into consideration. Seasonal changes have to be considered when arranging menu. Food items have to in line with the characteristics of a particular season. Changes in temperature can somewhat affect people's heat consumption as well as digestion and absorption of food. People's state of mind while eating can also be affected. Therefore, seasonal changes have to be taken into account when preparing family feast.

(2) Combination of both cold dishes and hot dishes and offering diverse flavors. Both cold dishes and hot dishes should appear at family feast. For either cold dish or hot dish, avoid offering two dishes (or more than two dishes) which have the same flavor. Normally, two cold dishes are offered for a feast of 2 to 4 guests, four cold dishes are offered for a feast of 5 to 7 guests while six dishes are offered for a feast of 8 to 10 guests.

(3) Collation of variety, color and nutrient content of food. When preparing family feast, both meat dishes and vegetables dishes can be served. Try different skills of cutting up meat and vegetables and food items can be diverse in color. One raw material can be made into different kinds of dishes such as "one fish for three dishes" and "one chicken for two dishes", an economical method which may create different styles of dishes.

(4) Knowing the dietary characteristics of guests. For example, people usually eat a lot during holidays and consequently, they may want high-quality food which is less oily and not readily available in normal times. Therefore, there should be less meat dishes on the table and each dish must have small amount of food. Habits and culinary preference of guests should also be taken into consideration. For example, people from Jiangsu, Zhejiang and Shanghai like sweet and light food while people from Sichuan and Guizhou prefer spicy and sour food.

(5) Eating freshly cooked food. In addition to some food items such as cold dishes, whole chicken and whole duck which are difficult to cook and have to be prepared in advance, stir-fried dishes taste best if freshly cooked. As food items and condiments contain various nutrient contents, taste and nutrient contents of food may change after being heated up. If food is eaten immediately after being removed from heat, the nutrient contents will be retained and the food will taste best. If food is heated up again after cooling off, the taste, aroma, flavor and appearance would all change and the dish would lose the original style.

A grand family feast focuses on order of courses served. The normal order is: cold dishes precede hot dishes; dishes precede dim sum; salty dishes precede sweet dishes; stir-fried dishes precede braised dishes; costly dishes (special flavor dishes) precede ordinary dishes; cooked dishes precede noodles and fruit. Specific order is as follows: ① Cold dishes for drinking wine. Cold dishes may be served as assorted dishes, four pairs of assorted dishes, four-three assorted dishes or an assorted dish surrounded with four, six or eight small single dish. ② Hot dishes cooked by using such techniques as smooth stir-frying, frying, deep-frying, saute, deep-frying and braising in order to diversify taste and appearance of dishes. ③ Main courses which are usually made with unbroken raw materials and served in large plate or large soup bowl. Cooking techniques employed include stewing, simmering, steaming and roasting. ④ Sweet dishes. Sweet dishes account for a small percentage of all dishes served. Normally, only one or two sweet dishes are offered. Cooking techniques may involve toffee, honey or such skills as deep-frying and steaming. Most sweet dishes are served hot. In summer, sweet dishes can be served as cold dishes. ⑤ Dim sum. Dim sum offered at family feast includes cakes, dumpling, noodles, crispy cake, steamed stuffed bun and jiaozi. The specific variety and texture of dim sum are determined by the scale of the family feast. ⑥ Soup. Different soup such as tomato egg soup and seaweed shrimp soup can be offered according to preference of guests. ⑦ Fruit.

If a family feast offers Western-style food, order of courses served is as follows: cold dishes; soup; hot dishes. Before the meal starts, aperitif and cocktail are offered first. Cold dishes are then first served. There are two ways of serving cold dishes: ① Placing all the cold dishes on the table before guests sit at the table; ② Serving cold dishes after guests sit at the table. Cold dishes include assorted dishes, salad, butter, jam and bread. Soup is served after cold dishes, and hot dishes are served after soup. If there are several courses of hot dish, make sure that there are both meat, fish, poultry meat, shrimps and vegetables. After guests have finished eating all the dishes, desserts and fruit are served which are followed by black tea and coffee. Milk and sugar should be offered at the same time so that some guests can take according to their needs.

After a dish is served on the table, the host should briefly introduce the flavor and style of the dish. If guests show interest in a dish, the host can also briefly introduce the cooking techniques employed. Normally, some guests at a table are senior in age and higher in status. The host should invite the senior in age or the guest of honor to first savor the food. If the guest is reluctant to savor the food first for humility's sake, the host can use the serving chopsticks to distribute food for guests. Food is usually first distributed to guests senior in age and then distributed according to seating order.

Dished should be served at appropriate time so that guests can enjoy their food and wine. When only one third of cold dishes is left, hot dishes can be served. Then guests can savor the food with wine as the accompaniment. After guests have drunk enough wine and main courses have been served, a course of sweet dish and sweet soup can be served so that guests can feel refreshed. Dishes, dim sum and soup are then served. After guests have dined and wined to satiety, the leftover food can be removed from the table. After cleaning the table, fruit is served signaling the end of the feast.

If invited to attend a feast, you should dress neatly, an act of respecting yourself and others. You should also be punctual. Except cocktail party, guests are usually asked to arrive 30 minutes before the banquet starts. It will not be considered impolite if you arrive just a few minutes before the banquet starts. However, it'll considered impolite and lack of respect for the host if you arrive late.

Say "hello" to host when you step into the host's room or banquet hall. In the meantime, shake hands with other guests or greet them with a smile or nod whether you know them or not. When guests senior in age arrives, stand up immediately and invite them to be seated and send your greetings. Behave in a dignified and refined manner in front of female guests.

Follow the suggestions of host or ushers regarding seating arrangement which is sometimes prearranged by the host. If there is no seating arrangement, remember the seat directly facing the entrance is seat of honor while seat directly facing the seat of honor across the table is seat of lower priority. Invite people of higher status, senior in age and female guests to be seated first, and then choose an appropriate seat for yourself.

After taking a seat, sit upright and keep feet under your own seat. Do not stretch out legs or shake legs continuously. Do not place elbows on the edge of table or place hand on the back of chair next to you. After sitting at the table, do not act as if there was no one else present. Do not stare at food on the table as if you can hardly wait to eat. Try to bring up conversations with those sitting around you.

Wear formal clothes when enjoying the food. Do not take off your coat before feast starts nor during the feast. Start eating only after the host gives a hint. Behave elegantly while eating and pick up food gently. Pick up food first on your own small plate and then use chopsticks to put food in mouth. Food should be eaten in small bites. Do not lean on table so as not to touch sitting beside you. Do not make sound while eating. If you need condiments placed in front of other guests, give a notice first. If the condiment is placed in an area out of your reach, ask for help politely. If you have to pick teeth during the meal, use hand or handkerchief to cover your mouth.

It's considered impolite to urge or even force other guests to drink during the meal (especially those who cannot hold liquor) and make them drunk.

If you have finished eating, do not leave the table if the feast is still ongoing. Leave the table only after the host and the guest of honor have finished eating and left the table.

Both tasty food and people's health are concerns of modern family feast. Most family feasts use serving chopsticks and serving spoons and a few family feasts offer individual servings. This hygienic mode can reduce waste, adapt to individual culinary preferences and make guests feel respected.

The ultimate purpose of family feast is to show your affection and loving care to your relatives and friends. For relatives and friends, what they value most from a joyous event or festive gathering is the "gathering" itself rather than the "food". As there are many topics to dwell upon during a family feast, no matter how delicious the food is, guests may feel cold-shouldered and lose interest if the host is just busy preparing the food and offering "service". Besides, in order to comply with the wishes of guests, the host must ask guests to eat more and at the same time, the host should not force guests to eat or drink. Just let the guests eat and drink as they think fit.

6.8 Graduation Feast

As a way of expressing gratitude to their teachers, graduation feast (or "further study feast") is the occasion when students invite their teachers for a feast upon their graduation from high school or college or when they have completed postgraduate study.

Hold a graduation feast is a voluntary act of student. As the purpose of holding a graduation feast is to express gratitude to teachers for their years' of academic training rather than a display of wealth and extravagance, family economic conditions should be taken into consideration when a student chooses the venue for the feast. Menu should be arranged according to the culinary preferences of each teacher and classmate invited to the feast. To increase joyous atmosphere and impress teachers and classmates, try to find something in common between the menu and teachers if the menu can be determined by the students. As dishes are the key to the success of a graduation feast, it's better to find out the culinary preferences of teachers beforehand. Try to invite every teacher from whom you have taken courses including your school counselor who has been closely associated with you.

At graduation feast, students should behave politely and avoid such impolite behavior as smoking and using bad language. To enliven the atmosphere, students can drink red wine or Chinese liquor as it's worth a celebration when students have passed entrance examination, achieved academic success or found a satisfying job. However, it's advisable not to drink too much wine. Normally, students will drink a toast to teachers, using the opportunity to express your innermost thoughts and feelings. As graduation feast is a meaningful part of a student's life journey, students may use the opportunity to take photos with teachers as a beautiful memory.

Graduation feast is of little significance if it's only about enjoying food. Students may design various activities at the feast to express gratitude to their teacher. Such activities include student expressing innermost thoughts and feelings, student chatting with teacher, and student representative and teacher representative gave a speech.

Common types of graduation feast include: classmates parting feast (inviting classmates to feast); great ambitions for child feast (inviting relatives and friends to feast); passing entrance examination feast (inviting teachers to feast).

The following is a reference for menu arrangement for graduation feast:

Bright Future — Classic assorted marinated dish

Studying Hard — Bitter gourd fried with sliced meat and fermented soya beans

Conscientiousness — Shredded eel fried with green pepper

Paying a debt of gratitude — Turtle fried with quail eggs

Light and cloud — Pumpkin and mung beans spareribs soup

High fighting spirit — Spiced crispy chicken

Marching forward bravely — Spareribs steamed with glutinous rice

Happy gathering — fruit platter

Student surpassing teacher — Braised seasonal vegetables in broth

Abundance year by year — Steamed mandarin fish

Honoring teacher and respecting his teaching is a traditional virtue in China. By holding a graduation feast, students may express their gratitude to teachers and know how to be thankful. However, an extravagant graduation feast may vulgarize the relationship between students and teachers and therefore it's not advisable to hold such a graduation feast.

6.9 Manchu Han Imperial Feast

Manchu Han imperial feast, reputed originating in the Qing Dynasty, the feast combines the best of Manchu cuisine and Han cuisine. It involves grill and hot pot from the Manchus and cooking methods of the Han people such as braising, deep-frying, stir-frying, quick-frying and stewing. The feast consists of a minimum of 108 dishes. Among the 108 dishes, there are 54 dishes from the South including 30 dishes from Jiangsu and Zhejiang, 12 dishes from Fujian and 12 dishes from Guangdong, and 54 dishes from the north including 12 dishes of the Manchu, 12 dishes from Beijing and 30 dishes from Shandong. During the reign of Emperor Qianlong (around 1764 AD), Li Dou in his *The Pleasure Boats of Yangzhou* describes a menu of a Manchu Han imperial feast, the earliest of its kind.

A Manchu Han imperial feast covers six banquets which last three days. All the six banquets are named after renowned banquets of the Qing imperial court. Before

Manchu Han imperial feast

sitting at the dining table, guests would burn two pairs of incense. Tea table and snacks are placed on top of table along with four kinds of fresh fruit, four kinds of dried fruit, four kinds of appreciation fruit and four kinds of preserves. After guests sit at the table, cold dishes are served first followed by hot dishes, main dishes and sweet dishes. Sets of pastel longevity dishware are used for dining accompanied by silverware, both are gorgeous and magnificent. Amid the refined and solemn atmosphere for dining, musicians would play music as an accompaniment and guests would enjoy food and music at the same time.

Manchu Han imperial feast features both dishes of imperial household and dishes of local cuisines. The feast offers choice food, follow strict etiquette and use a wide range of fine ingredients including exotic ones from mountains and seas. Employing exquisite cooking skills, the feast offers dishes of style of different regions, primarily dishes from the northeast, Shandong, Beijing, Jiangsu and Zhejiang. Most of the so-called costly dishes at Manchu Han imperial feast are made from local specialties (or products) from Heilongjiang. Dishes from Fujian and Guangdong cuisines also appeared at the feast at later stage. According to Baidupedia, the six banquets of Manchu Han imperial feast include:

1) Banquet for Mongolian vassals

The banquet is a royal feast held by the Qing emperor for Mongolian vassals who had united with the Qing royal house through marriage. The banquet was usually held at the Hall of Uprightness and Brightness and officials of the first rank and the second rank were also asked to attend. Emperors of Qing Dynasty gave high priority to the banquet and held it on an annual basis. The Mongolian vassals attending the banquet viewed the feast as a blessing and greatly cherished dishes awarded by the emperor during the banquet.

2) Banquet for ministers

Banquet for ministers was usually held on the 16th day of the first lunar month. The emperor would select grand secretary and those among the nine ministers who had rendered meritorious service to attend the banquet, a great honor for those selected. The banquet was usually held at "The Hall of the Three Selflessness" and followed etiquette of imperial clan banquet. Sitting on a high back chair, those attending the banquet would compose poems and drink wine. The banquet was held at an annual basis and the Mongolian nobility were also asked to attend. Emperors tried to win over his ministers through this patronizing act. The banquet was also a symbol of the status and honor of those high-ranking officials who were able to attend the banquet.

3) Longevity banquet

As one of the major banquets of the royal house, longevity banquet was the birthday banquet for the Qing emperor. Concubines, princes and civil and military officials considered it an honor to be able to offer birthday congratulations and present birthday gifts to the emperor. Countless dishes of choice food were offered at the banquet. The celebration would be more magnificent and on a larger scale if it's the significant birthday of the emperor. Everything imaginable was available at the banquet such as fine clothes and jewelry, splendid decorations and furnishings, dances and music, and choice food. The 10th day of the 10th lunar month during the 20th year reign of Emperor Guangxu was the 60th birthday of Empress Dowager Ci Xi. To celebrate her birthday, an imperial edict was issued during the 18th year reign of Emperor Guangxu and the birthday banquet started several months before the birthday date. More than 29,170 glazed bowls, saucers and plates bearing the Chinese characters for "Living Forever" and auspicious patterns had been made in Jiangxi. Nearly 10 million tales of silver had been spent on the celebration, which was unprecedented in Chinese history.

4) Nine white banquet

Nine white banquet was initiated during the reign of Emperor Kangxi. When Emperor Kangxi designated four tribes outside Mongolia, these tribes paid "nine white" as tribute to the emperor as a way of showing loyalty. As a sign of pledge, these "nine white" included one white camel and eight white horses. After receiving tribute from these tribes, the emperor would hold an imperial feast to entertain the envoys. Called "nine white banquet", the banquet was held on an annual basis.

5) Festival and seasonal banquets

Festival and seasonal banquets were feast held by the Qing royal house according to set rules at a particular festival or in a particular season. These banquets include lunar New Year's Day banquet, lunar New Year's Day gathering banquet, spring ploughing banquet, Dragon Boat Festival banquet, Qiqiao Festival banquet, Mid-Autumn Festival banquet, Double Ninth Festival banquet, Winter Solstice Festival banquet, New Year's Eve banquet. Although the Manchus had their own culinary customs, they adopted many culinary customs of the Han people after conquering China for the purpose of integrating cultures of the Manchu and the Han and strengthening their rule over the country. Due to the unique status of the imperial court, more detailed culinary rules had been formulated. In the meantime, its culinary preferences were also closely associated with those of the common people and different regions. As a result, food items such as laba congee, yuanxiao, zongzi, ice bowl, realgar wine, Chongyang cake, Qiqiao cake and moon cake were also available in the Qing imperial court.

6) Banquet for 1,000 Elders

Initiated during the reign of Emperor Kangxi and grew during the reign of Emperor Qianlong, banquet for 1,000 elders was the largest feast held by the Qing imperial court which had the largest number of attendees. The first banquet for 1,000 elders was held in Yangchun Garden during the 52nd year of reign of Emperor Kangxi. Emperor Kangxi composed a poem titled "Banquet for 1,000 Elders" at the banquet; hence the banquet was called the "banquet for 1,000 elders". The banquet was again held in the Palace of Heavenly Purity during the 50th year of reign of Emperor Qianlong with 3,000 attendees. Another such banquet was held at the Hall of Imperial Supremacy of the Palace of Tranquil Longevity during the first lunar month of the first year of reign of Emperor Jiaqing with 3,056 attendees and more than 3,000 poems composed at the banquet. Later generations described the banquet as "an unprecedented event in history" which demonstrated "abundant grace and proper etiquette".

The dishes of the Manchu Han Imperial Feast are based on that time, but there are quite a few changes now. Therefore, the specific dishes will not be listed.

Today, although we don't have the opportunity to sample all dishes from the Manchu Han Imperial Feast, we can choose some from the menu and learn about their cooking methods from books and online information so as to enjoy choice food from different regions of the country and thereby appreciate the splendid culinary culture of China.

In the 21st century, with improvement of living standards and transformation of consumption concepts, Chinese people have started to favor novel, nutritious and hygienic dishes. Their pursuit has promoted the advancement of banquet culture which is entering a new stage. Traditional banquet has undergone improvement while new style of banquet has been created. At the same time, the Western style has been introduced leading to fusion between the Chinese style of banquet and the Western style of banquet. As a treasure inherited from ancestors of the Chinese nation, Chinese style of banquet is an epitome of the tranquility, beauty and novelty of the oriental culture. As we continue to learn from our traditional culture, we should inherit our heritage with a critical thinking and make more progress through innovation!

Bibliography

[1] Loving Family Cuisine Food Culture Studio. *New Home Cooking Recipes* 2[M]. Qingdao: Qingdao Publishing House, 2009.

[2] Chen Guangxin. *Collection of Writings on Chinese Banquets*[M]. Qingdao: Qingdao Publishing House, 2001.

[3] Chu Dan. *Calligraphy*[M]. Hefei: Huangshan Publishing House, 2011.

[4] Editorial Committee of *Collection of Super Value. Complete Works of Chinese Folk Customs* [M]. Beijing: China Pictorial Press, 2012.

[5] Du Daming. *Chinese Culinary Culture*[M]. Shanghai: Fudan University Press, 2011.

[6] Duli, Yaohui. *Chinese Culinary Culture*[M]. Beijing: Tourism Education Press, 2016.

[7] Hu Zhiqiang. *Most Interesting Facts About Folk Customs*[M]. Beijing: China Textile & Apparel Press, 2012.

[8] Ke Ling. *Chinese Folk Customs and Culture*[M]. Beijing: Peking University Press, 2017.

[9] Li Lin. *Introduction to Tea and Wine Culture*[M]. Taiyuan: Beiyue Literature and Art Publishing House, 2010.

[10] Liu Yi, Rui Hong. *Living In Jungles and Mountains: Hunting and Gathering Life of Ethnic Groups in Yunnan*[M]. Kunming: Yunnan Education Publishing House, 2000.

[11] Liu Yong. *Chinese Wine (Chinese and English Version)* [M]. Hefei: Huangshan Publishing House, 2012.

[12] Ma Jianying. *History of Chinese Culinary Culture*[M]. Shanghai: Fudan University Press, 2008.

[13] Kong Jianmin. *Collection on Life Tips*[M]. Nanjing: Jiangsu Science and Technology Press, 2003.

[14] Qiao Jiaojiao. *Chinese Food*[M]. Hefei: Huangshan Publishing House, 2014.

[15] Qu Mingan, Qin Ying. *History of Chinese Food and Entertainment*[M]. Shanghai: Shanghai Classics Publishing House, 2011.

[16] Wang Xuetai. *A Brief History of Chinese Culinary Culture* [M]. Beijing: Zhonghua Book Company, 2010.

[17] Wu Peng et al. *Chinese Culinary Culture*[M]. Beijing: Chemical Industry Press, 2013.

[18] Ye Changjian. *Chinese Culinary Culture* [M]. Beijing: Beijing Institute of Technology Press, 2011.

[19] Xie Dingyuan. *Chinese Culinary Culture*[M]. Hangzhou: Zhejiang University Press, 2008.

[20] Xiao Fan. *Encyclopedia of Chinese Cooking* [M]. Beijing: Encyclopedia of China Publishing House, 1992.

[21] Yao Weijun, Liu Pubing, Ju Mingku. *History of Classics on Chinese Food* [M]. Shanghai: Shanghai Classics Publishing House, 2011.

[22] Ye Lang and Zhu Liangzhi (author), Cathy (translator). *Chinese Culture Reader (German Version): Blick auf die Chinesische Kultur* [M]. Beijing: Foreign Language Teaching and Research Press, 2014.

[23] Yu Weijie. *Food Ingredients History of China* [M]. Shanghai: Shanghai Classics Publishing House, 2011.

[24] Yu Ye. *Chinese Tea (Chinese and English Version)* [M]. Hefei: Huangshan Publishing House, 2011.

[25] Zhang Enlai. *Family Kitchen Know-all* [M]. Nanjing: Jiangsu Science and Technology Press, 2004.

[26] Zhang Jingming. *A Study on Culinary Culture of Nomadic People in Northern China* [M]. Beijing: Cultural Relics Press, 2008.

[27] Zhang Jingming, Wang Yanqing. *History of Chinese Eating Utensils* [M]. Shanghai: Shanghai Classics Publishing House, 2011.

[28] Zhao Fanyu, Shui Zhongyu. *33 Etiquette Lessons Before 30* [M]. Shanghai: Lixin Accounting Publishing House, 2010.

[29] Zhao Jianmin, Jin Hongxia. *Introduction to Chinese Culinary Culture* [M]. Beijing: China Light Industry Press, 2011.

[30] Zhao Rongguang. *Chinese Culinary Culture* [M]. Beijing: Zhonghua Book Company, 2012.

[31] Zhao Rongguang. *Introduction to Culinary Culture of China* [M]. Beijing: Higher Education Press, 2003.

[32] Zhao Rongguang. *Chinese Wine Culture* [M]. Beijing: Zhonghua Book Company, 2012.

[33] Zhuge Wen. *Understanding 5,000 Years of Chinese Folk Customs in Three Days* [M]. Beijing: China Legal Publishing House, 2014.

[34] Chen Juan. Impact of Geographical Conditions on Chinese Culinary Culture [J]. *Journal of Fujian Education Institute*, 2003(4).

[35] Hu Xiaoping. Chongyang Wine [J]. *International Finance*, 1999(11).

[36] Hou Zhongming. Chinese Calligraphy and Wine Culture [J]. *Journal of Sichuan University of Arts and Science*, 1999(1).

[37] Li Jun. A Comparative Study of Chinese and Korean Traditional Festival Food Culture — Spring Festival and Mid-Autumn Festival [J]. *Asia-Pacific Education*, 2016 (29).

[38] Liu Changhao, Li Xiumin. Problems of Chinese Traditional Snacks and Correction Measures [J]. *Business Journal*, 2018(7).

[39] Lǚ Xiaomin, Ding Xiao, Dai Yangyong. Development and Background of Chinese Great Eight Traditions [J]. *Food and Nutrition in China*, 2009(10).

[40] Jiang Zhiying. Chinese Great Eight Traditions and the Ninth Great Tradition [J]. *Literature and*

History, 2013(5).

[41] Manlai. Varieties and Classification of Beijing Snacks [J]. *Economic and Trade Update*, 2012(5).

[42] Pang Qianlin, Lin Hai, Ruan Liuqing and Li Ximing. Chinese Rice Culture and Modern Achievements[J]. *China Rice*, 2004(3).

[43] Qin Dadong. A Brief History of Dark Tea[J]. *Journal of Tea Business*, 1983(6).

[44] Wan Jianzhong. History of Chinese Regional Cuisines[J]. *People's Weekly*, 2016(6).

[45] Wang Wenhui, Tong Wei, Jia Xiaohui. Frozen Pear Production History, Industry Situation and Problem Analysis[J]. *Storage and Process*, 2015(11).

[46] Yang Zhongjian. Shanghai Benbang Cuisine[J]. *Sichuan Cuisine*, 2014(11).

[47] Yang Shuqiong. Chinese Character for "Wine" and "Wine Culture" [J]. *Journal of Inner Mongolia Radio & TV University*, 2017(6).

[48] Yang Lin. Origin of Mid-Autumn Festival[J]. *Tracing Roots*, 1997(8).

[49] Yang Jian, Sun Daqing. Study on Microbial Diversity In Leavened Dough of Northeast Sticky Steamed Bun Stuffed with Sweetened Bean Paste [J]. *Journal of Heilongjiang Bayi Agricultural University*, 2015(10).

[50] Yang Ze. Origin of Mid-Autumn Festival[J]. *Yangtze and Huai Rivers*, 2004(9).

[51] Zhou Wang. Study on Characteristics of Regional Snack Culture of Ethnic Minorities in Guangxi, Yunnan and Guizhou in Southwest China[J]. *Research on Culinary Culture*, 2005(3).

[52] Zhang Yan, Zhang Cheng. Study on Culinary Culture of Ethnic Minorities in Northeast China [J]. *Cultural Highland*, 2014(1).

[53] Liu Hui. *Study on Benbang Cuisine Culture Transmission at Shanghai Old Restaurants* [D]. Shanghai: East China Normal University, 2015.